James E. H. Gordon

Physical Treatise on Electricity and Magnetism

Volume 1, Second Edition

James E. H. Gordon

Physical Treatise on Electricity and Magnetism
Volume 1, Second Edition

ISBN/EAN: 9783337406097

Printed in Europe, USA, Canada, Australia, Japan

Cover: Foto ©berggeist007 / pixelio.de

More available books at **www.hansebooks.com**

A PHYSICAL TREATISE

ON

ELECTRICITY AND MAGNETISM.

VOL. I.

LONDON :
GILBERT AND RIVINGTON, LIMITED,
ST. JOHN'S SQUARE.

A PHYSICAL TREATISE

ON

ELECTRICITY AND MAGNETISM.

BY

J. E. H. GORDON, B.A. Camb.

MEMBER OF THE INTERNATIONAL CONGRESS OF ELECTRICIANS, PARIS, 1881;
MANAGER OF THE ELECTRIC LIGHT DEPARTMENT OF THE TELEGRAPH
CONSTRUCTION AND MAINTENANCE COMPANY.

SECOND EDITION.

REVISED, RE-ARRANGED, AND ENLARGED.

IN TWO VOLUMES.

VOL. I.

London:

SAMPSON LOW, MARSTON, SEARLE, & RIVINGTON,

CROWN BUILDINGS, 188, FLEET STREET.

1883.

PREFACE TO THE FIRST EDITION.

My object in this treatise has been to give a complete account of such portions of Electrical Science as I am acquainted with, and at the same time to regard them from a physical as distinguished from a mathematical point of view.

I have throughout endeavoured, as much as I am able, to connect the various phenomena described with the hypothesis adopted by Newton, Faraday, and Maxwell, namely—that there is no such thing as "action at a distance," but that all electrical actions are transmitted from place to place by strains of some continuous medium filling the space between.

In order that the book may be more useful, I have entered very fully into the actual experimental details of most of the investigations mentioned, partly in order that any one wishing to continue the experimental investigation of the subject may find in the book accounts of what has already been done, and partly that students may see that electrical phenomena have a real existence, and are not merely abstractions or results of mathematical analysis.

I may add that, for the understanding of the text of the work, no mathematical knowledge is required except that of algebra up to simple equations. There are, however, a few mathematical foot-notes and appendices.

By far the greater number of the illustrations in the book are original, and have been engraved under the directions of Mr. Cooper, from drawings of the most recent instruments now in actual use, made by Mr. Cole and Mr. Webster, at Kew Observatory, at Cambridge, at King's College London, at several private laboratories, and at the factories of the principal instrument makers; and I must now express my sincere thanks to these three gentlemen for the skill and care displayed by them throughout the whole of a long and laborious task.

I also must express my thanks for the loan of blocks and plates to the Councils of the Royal Society, of the British Association, and of the Royal Irish Academy, to the Committee of Kew Observatory, to Mr. **Wm.** Spottiswoode (President of the Royal Society), Sir Wm. Thomson, Mr. **De La Rue,** Mr. Crookes, Prof. Tyndall, the late Prof. Clerk-Maxwell, Dr. Kerr, M. Planté, Prof. Röntgen, Prof. **Hughes,** Prof. **Jellett,** Mr. Ayrton, Mr. Perry, Messrs. Elliott Bros., and Mr. Apps.

I have to return most hearty thanks to many friends for invaluable assistance with the proof-sheets and **MSS.**

Mr. Ayrton has read the whole of Part **I. in** proof, and most of Part **III.** He has paid special attention **to** the chapter on "Contact **Electricity."**

Mr. Whipple, Superintendent of **Kew** Observatory, has read the whole of Part II. (Magnetism) in MS., in slips and in sheets; **he has** had several tables prepared for me at **the** Observatory, and has superintended **the drawings of the various** magnetic instruments which were made at Kew by Mr. **Webster.**

Prof. Everett has read all those portions of the book which relate to the theory of Units, namely, Chapters VIII., XXI., and XLVI.

Prof. Tyndall **has read the** MS. **of** the chapter on Diamagnetism.

Mr. Spottiswoode has read the proof of all the chapters about the Induction **Coil, and** about **Electrical Discharges,** i. e. Chapters XXXIX. and XLI. to XLIII. **Mr. Moulton** has also read the chapter on **"The Sensitive State."**

Mr. Kieser, of the firm **of Elliott Bros., has read** the description of the various forms of Resistance Boxes.

Dr. Kerr has read Chapters **XLIX. and** L., in which an account of his discoveries is given.

Prof. W. G. Adams has also given me some valuable assistance.

Prof. Boltzmann, of Vienna, has had the great kindness to **write** out for me the mathematical theory (not before published) **of** his experiments on Specific Inductive Capacity. The appendix to Chapter XI. is substantially **a translation of** his letter.

Mr. J. G. Butcher and Mr. T. O. **Harding** have also given me much assistance with the appendix **to** Chapter XI. and with **some** of the mathematical foot-notes.

Holmwood, Dorking, April 17, 1880.

PREFACE TO THE SECOND EDITION.

THE demand for a Second Edition of this book came upon me a year ago, much sooner than I expected it. As my life is no longer that of a student, but is fully occupied by engineering work, time for writing has been scarce, and this book has been long out of print. Now that it is ready for the press, I can only, for the same reason, bespeak indulgence for it.

The principal changes introduced into the new edition are, first, correction of a few errors; second, a re-arrangement of some of the chapters; third, the introduction of new matter, so as to bring the book as much as possible up to date. The principal new matter introduced has been Chapter XXXIII., which contains an account of Mr. Tribe's electrolytic experiments, and which is illustrated by facsimiles of the beautiful copper deposits on silver plates prepared by him; and Chapters XXXVI., XXXVII., which deal with the wonderful analogies between mechanical vibrations and electric and magnetic phenomena observed by Professor Bjerknes and by Mr. Stroh. The chapter on Professor Bjerknes' experiments is illustrated by plates photographed from tracings most kindly prepared for me by Professor Bjerknes and by his son, Mr. Vilhelm Bjerknes.

There are also accounts of some new experiments by Mr. De La Rue, M. H. Becquerel, and others.

The subject of Electric Lighting has now become so important that I felt it would be impossible to treat it satisfactorily in the space available in a treatise on general electrical science. I have therefore cut out the short chapter on it which appeared in the First Edition, and am reserving what I have to say for a work specially devoted to electric lighting, which is now nearly complete, and which I hope shortly to publish.

28, *Collingham Place, S.W., London.*
June 1st, 1883.

CONTENTS OF VOL. I.

Part I.

ELECTRO-STATICS.

CHAPTER I.

PRELIMINARY.

CHAPTER II.

ELECTRICAL MACHINES.

CHAPTER III.

QUANTITIES OF ELECTRICITY.

CHAPTER IV.

ELECTRIC FORCE.

CHAPTER V.

CHAPTER VI.

POTENTIAL.

CHAPTER VII.

ELECTROMETERS.

CHAPTER VIII.

ON THE THEORY OF ABSOLUTE MEASUREMENT.

Contents.

CHAPTER IX.

ABSOLUTE ELECTROMETERS.

CHAPTER X.

THE LEYDEN JAR.

CHAPTER XI.

SPECIFIC INDUCTIVE CAPACITY.

Contents.

Part II.

MAGNETISM.

CHAPTER XII.

MAGNETISM—PRELIMINARY EXPERIMENTS.

Contents.

CHAPTER XIII.

TERRESTRIAL MAGNETISM—HISTORICAL NOTES.

CHAPTER XIV.

TERRESTRIAL MAGNETISM—MODERN EXPERIMENTAL METHODS.

CHAPTER XV.

SELF-RECORDING MAGNETOMETERS.

CHAPTER XVI.

OBSERVATIONS ON TERRESTRIAL MAGNETISM.

CHAPTER XVII.

EARTH CURRENTS.

Part III.

ELECTRO-KINETICS.

CHAPTER XVIII.

THE VOLTAIC BATTERY.

CHAPTER XIX.

ELECTRICITY OF CONTACT.

CHAPTER XX.

ACTIONS OF CURRENTS ON MAGNETS—COMMUTATORS—GALVANOMETERS.

CHAPTER XXI.

ELECTRIC RESISTANCE AND ELECTRO-MAGNETIC UNITS.

CHAPTER XXII.

EXPERIMENTAL MEASUREMENT OF RESISTANCES.

CHAPTER XXIII.

ELECTRO-MAGNETISM—PRELIMINARY NOTES.

CHAPTER XXIV.

THE TELEPHONE AND MICROPHONE.

CHAPTER XXV.

CHAPTER XXVI.

CHAPTER XXVII.

BRITISH ASSOCIATION UNIT OF RESISTANCE.

CHAPTER XXVIII.

MUTUAL ACTION OF CURRENTS ON EACH OTHER AND BETWEEN CURRENTS AND MAGNETS.

CHAPTER XXIX.

RELATION BETWEEN VARIATION OF POTENTIAL AND STRENGTH OF CURRENT

CHAPTER XXX.

CURRENTS PRODUCED BY INDUCTION ON CLOSING AND BREAKING THE CIRCUIT.

LIST OF PLATES IN VOL. I.

PART I.

ELECTRO-STATICS.

A PHYSICAL TREATISE

ON,

ELECTRICITY AND MAGNETISM.

Part I.

ELECTRO-STATICS.[*]

CHAPTER I.

PRELIMINARY.

WE have as yet no conception of electricity apart from the electrified body; we have no experience of its independent existence, and therefore we will commence our study of the science by considering the "properties of electrified bodies."

Properties of Electrified Bodies.

DEFINITIONS.

When a body[†] shows certain properties, it is said to be *electrified*. These properties are:—

(1.) The power of **attracting** or repelling other electrified bodies.

(2.) The power of inducing similar properties on neighbouring bodies.

(3.) The power of attracting light unelectrified bodies,[‡] and

(4.) When very strongly electrified, the power of giving off sparks.

Bodies can be *electrified* by various methods.

One method is by "**friction.**" If a rod of glass or a rod of

---

[*] See page 28. [†] Not being iron or steel.

[‡] (3) will be shown to be a consequence of (1) and (2)

52

sealing-wax be rubbed with a silk handkerchief, it will be electrified.

It is found that the electrification of glass differs from that of sealing-wax.

EXPERIMENTS.

In the following experiments we shall require two pieces of stout glass tube, such as is used for barometers, each about 14 inches long and ½ to ¾-inch diameter, two rods of coarse sealing-wax of about the same size, and a silk handkerchief. We shall also require a wooden stand (fig. 1), consisting of two uprights about 18 inches apart, connected by a cross-bar at the height of about two feet; from the centre of the cross-bar a wire stirrup must be suspended by a silk thread about 18 inches long.

Fig. 1.

The wire stirrup must be of such a shape that when one of the glass or sealing-wax rods is placed in it, it will be able to turn freely in a horizontal plane like a compass needle.

The experiments must be done in a warm dry room, and the rods and handkerchief must all be warmed by the fire for some time before-hand.

Now rub two of the rods briskly with the handkerchief,* place one in the stirrup and hold the other near the rubbed end of it. It will be observed that, if the two glass rods or the two sealing-wax rods be used, there will be a strong *repulsion*, and that if one glass and one sealing-wax be taken, there will be an equally strong *attraction*.

That is to say, *similar electrifications repel each other, unlike attract.*

It is found that all electrification is of one of two kinds; it is all either the same as that produced by rubbing glass, or as that produced by rubbing sealing-wax.

It is also found that, if equal quantities of the electricity of glass and the electricity of sealing-wax be added together, they neutralize each other.

---

* This experiment succeeds better if two experimenters each with a handkerchief rub the rods simultaneously.

(+) AND (−) ELECTRICITY.

For this reason the two electricities are called respectively "positive" and "negative," written (+) and (−).

The electricity of glass has been arbitrarily chosen to be called (+).

We know from elementary algebra that if a quantity with the (+) sign be added to an equal quantity with the (−) sign, the result is zero, and therefore, the fact that the result of adding equal quantities of the electricity of glass and the electricity of sealing-wax, is zero, enables us to treat them as ordinary algebraical quantities with (+) and (−) signs.

ATTRACTION OF LIGHT BODIES.

If the stirrup be removed from the frame (fig. 1), and a feather attached to the silk instead of it, it will be found that all the electrified rods, whether glass or sealing-wax, attract the feather.

ELECTRIFICATION OF THE RUBBER.

When a rod is electrified by friction, it is found that the rubber is electrified with an electrification opposite to that of the rod.

This is not a very easy thing to show experimentally, owing to the rapid escape of the electricity of the rubber through the hand.

If, however, the silk be protected by holding it in a bit of sheet india-rubber, it will generally be possible to observe its attractive effect on the feather, and its attractive and repulsive effects on the electrified glass and sealing-wax, when the rods of them are placed in the stirrup.

The quantity of electricity of one kind produced on the rubber is exactly equal to the quantity of the other kind produced on the rod.

If the rod be vigorously rubbed with the handkerchief and the latter not removed, then, if the rod and handkerchief together be brought near the suspended feather, no attraction will take place.

This shows that the electrification of the rod is exactly neutralized by the opposite electrification of the handkerchief, and, therefore, that the two quantities of electrification must be exactly equal.*

TRANSFER OF ELECTRIFICATION.

If an electrified body **be made** to touch one not previously electrified, it is found that the one **loses a part of its** electrification and the other gains electrification.

We then say that part of the *electricity* of the charged body has been transferred to the other body. The quantity transferred depends on the sizes and shapes of the two **bodies.** If they are both spheres, the electricity will divide **between them in** direct **proportion** to their radii.

Experiment.—Attract the **feather by one of** the **rubbed** rods **and let** it touch it. It will **be** found to cling **to** it for awhile. Turn the rod gently round, so **as to make** different portions of its surface touch the feather. In the **course** of half a minute or so, the feather will fly away from **the rod,** and, on bringing the latter near it, it will be repelled. This is explained by the fact that, when **the feather touched the rod, a** portion of the electrification of the latter was transferred to it, and it is **repelled** because **it is charged with** the **same** electrification as that on the rod.

CONDUCTORS AND INSULATORS.

If one end **of a rod of** sealing-wax be electrified, it keeps its electrification **for a considerable** time in dry **air,** and the other end remains unelectrified.

If, on the other hand, one end of a rod of metal be electrified, the electrification at once distributes itself all over the rod ; and if one **end of** the rod be placed on the ground, the electrification **is rapidly lost.**

Substances like the sealing-wax, in which electricity cannot **move** freely, are called "*Insulators,*" those like metals, in which **it can move** freely, are called "*Conductors.*"

Silver is the best conductor known, and dry air the best **insu-**lator, but no substances are quite perfect either way.

So many substances exist whose conducting and insulating

* The experiment must **be** performed **with a** feather, and not with the suspended electrified rod, as the suspended rod would attract any unelectrified body brought near it in the same way as the electrified rod attracts the feather.

powers come **between** those of the above, that there can hardly be said to **be any** definite line of demarcation between conductors and insulators.

We may, however, for all practical purposes, consider the metals as conductors, and air, glass, **ebonite**, sealing-wax, paraffin-wax, and silk, as insulators.

THE GOLD-LEAF ELECTROSCOPE.

This is a more delicate apparatus than the suspended feather for detecting small quantities of electrification.

It shows the presence of electricity by the divergence of two gold-leaves hung close together. It is described on page 32.

ELECTRIFICATION BY INDUCTION.

When a charged body is placed near another, but not in contact with it, the second body becomes electrified, and remains electrified as long as the charged body is in its neighbourhood. As soon as the charged body is removed, all signs of electrification in the other disappear.

This kind of electrification is called "*Electrification by induction.*"

Experiment.—Rub one of the rods and hold it near to, but not in contact with, the plate of the electroscope. The **leaves at** once diverge, and remain diverged as long as the rod remains in position. On removing the rod they collapse.

The experiment still succeeds if a plate of glass, gutta-percha, paraffin-wax, *or any other non-conductor*, be placed between the electroscope and the sealing-wax.

This shows that the charged rod induces a charge of electricity in the plate and gold-leaves of the electroscope, *the action taking place at a distance and across the intermediate insulator.*

METAL SCREENS.

It is found that, if a large metal screen, which is connected to the earth to allow electricity to escape from it, be placed between the sealing-wax and the electroscope, no induction takes place. Similarly the induction can be stopped by covering the electroscope with a metal or wire gauze cage connected to the earth.[*]

This fact is very important in many experiments, as, by being

* See page 7

enclosed in metal tubes and cases, wires and instruments can be protected from the induction of neighbouring charged bodies.

NATURE OF THE INDUCED CHARGE.

It is found that the induced charge invariably consists of equal quantities of positive and negative electricity.

The electricity of the opposite kind to that of the inducing body is on the side next to it and, in a conductor,[*] the electricity of the same kind is on the further side.

Experiment.—Place a metal conductor on a glass stand near a strongly electrified body (such as the conductor of an electric machine, fig. 2). It will become charged by induction. Place two gold-leaf electroscopes at a sufficient distance from it for them not to be affected directly by the charged body.

Let us now explore the metal conductor by means of a " proof-plane."

A " proof-plane " is a metal disc about 2 inches diameter attached to an insulating handle.

By means of it, a portion of the charge of any part of the conductor can be carried to an electroscope; for on touching the conductor with the proof-plane, a portion of the charge passes to it, and, on touching the electroscope with the proof-plane, part of the charge which the latter has acquired passes to the former.

Let us first touch the middle of the conductor with the proof-plane, and carry it to one of the electroscopes. No effect will be produced. This shows that all the induced charge is at the ends.

Let us now touch the two ends respectively with two proof-planes, and then convey the charges obtained to the electroscopes respectively.

Both sets of leaves will diverge, showing that both ends of the conductor were charged.

Let us now connect the two electroscopes by a fine wire held in an insulating handle. Both sets of leaves will collapse, show-ing that the electricities from the two ends were opposite in kind and have neutralized each other when allowed to combine.

---

[*] In non-conductors the distribution of the similar electricity is more complicated.

Now charge a small pith ball, suspended by a silk thread, with the same electricity as that of the inducing body, and, bring it near to the far end of the conductor.

It will be repelled, showing that the electricity of the far end is of the same kind as that of the inducing body, and, therefore, that of the near end of the opposite kind.

Now remove the inducing body; all signs of electrification disappear from the conductor, showing that the charges of positive and negative electricity were exactly equal in amount, as they have neutralized each other when set free and allowed to combine.

ATTRACTION OF LIGHT BODIES EXPLAINED.

We now see why a charged body attracts light bodies not previously charged. It induces a charge opposite to its own on the near side, and one similar to its own on the far side. The near side is attracted and the far side repelled; but, owing to the smaller distance of the attracted side, the attraction is stronger than the repulsion and the body as a whole is attracted.

EARTH CONNECTION.

In certain experiments we wish to study only one portion of the induced charge, namely, that induced on the end of the conductor which is nearest to the charged body.

The charge on the far end can be removed to any distance by sufficiently lengthening the conductor.

Instead of making the metal conductor of great length, we may carry a wire from it to the nearest water-pipe; the whole world then becomes part of the conductor. The "near end" is that of the metal conductor under examination; and the "far end" may be considered to be somewhere in Australia.

If a conductor "connected to earth" be acted on inductively by a positively charged body, the whole charge on it will be negative—that is, all the parts near to the charged body will be negative, and no charge can be detected on the other parts, for they will all be in the same condition as the *middle* of the conductor in the experiment described on page 6.

In speaking of this phenomenon we sometimes say that, when the conductor is connected to earth, the induced electricity of the "far end" escapes to earth. This must merely be considered as an abbreviated way of stating what occurs.

CHAPTER II.

ELECTRICAL MACHINES.

FOR the production of large quantities of statical electricity, various machines are used, as the rubbing of a glass or sealing-wax rod with a handkerchief, only produces a small amount.

The principal machines used are those which we are about to describe, namely, the " Friction machine," the " Holtz machine," and the " Electrophorus."

THE FRICTION MACHINE.

The Friction machine is essentially an apparatus for conveniently rubbing glass and silk together, and collecting the electricity produced.

It is made in several forms. *The glass plate machine* (fig. 2)

Fig. 2.

consists of a circular plate of glass, which can be revolved on an axis by turning a handle. **Two pairs of cushions** covered with silk press against it near its edges. The cushions are connected with the ground, so that the negative electrification produced on them may disappear.

The glass becomes highly charged positively. **A massive** brass conductor is fixed, insulated, to the stand of the machine, and two branches of it project so as almost to touch the glass just after it leaves the rubbers.

These branches are furnished with a row of points to **conduct** the electricity from the glass.* On turning the handle, **the conductor** becomes highly charged with positive electricity. **On** placing **the knuckle near the conductor, sparks can be** obtained.

A machine whose plate is 2 feet in diameter should, when in good order, give sparks nearly 2 inches long.

Glass machines are also made in a cylinder form.

The objection to glass machines is their **liability to break,†** and the impossibility **of** making **them** work in **damp weather,** owing to the way in which moisture **collects upon the glass.**

Plate machines, which do not **suffer from these defects,** are sometimes constructed of ebonite.

At the Cavendish laboratory at Cambridge, an elaborate machine for drying the **air is** erected in the electrical room.

THE HOLTZ MACHINE.‡

This machine works not by friction, but **by induction.**

* See page 22.
† When **a machine is placed near** the fire **to warm** it before use, it should have the edge **of its plate turned to** the fire, **as** the plate is then less liable to crack than if **placed with its side** exposed to **the** heat.

‡ THEORY OF THE ACTION OF THE HOLTZ MACHINE.
Holtz machines have been constructed in a cylindrical form. In this case the windows are at opposite sides of a circular fixed cylinder, inside which the revolving cylinder turns.
The combs are again inside the moving cylinder.
The action of this machine is precisely the same as that of the plate form, and though it is not a useful machine for practical purposes, we will use it to explain the theory, as the diagram we shall require is more easily drawn for it than for the plate machine.
The following explanation of the action is due to M. Mascart :—
The theory of the machine in its various forms will be found very **fully and** excellently discussed in his *Electricité Statique*, T. ii. p. 278.

It consists of two parallel circular glass plates placed near to each other (fig. 3). One, D, which is solid, can be revolved rapidly by means of a multiplying wheel; the other, D', is fixed, and has in it three holes; one circular, at the centre, to allow the axis of the revolving plate to pass through—and two openings, F G, called "windows," at opposite sides and ends of a diameter, usually horizontal. Fixed at the horizontal edges of these windows are pieces of paper with somewhat blunt tongue-shaped points, *a a, b b,* projecting through the windows, so as to lightly touch the revolving plate. Near the other side of the revolving plate and opposite to the windows are fixed metal "combs" connected re-.

Fig. 3.

spectively to the two discharging rods, between which the spark is obtained.

To work the machine, one of the paper armatures is charged by means of a piece of rubbed sealing-wax or a small friction machine, and the discharging rods, N N', P P', are placed in contact with each other. After the machine has been set in motion (which

The cylindrical form of the machine being represented by fig. 4 (in which the outer cylinder is not shown, but only the paper armatures which it supports) let us suppose that the armature A is charged with (−) electricity, that the movable cylinder turns in the direction of the arrows, and

must be in such a direction that the plate travels *towards* the projecting tongues), the discharging rods are separated, and a that the balls P and N are in contact, so that the conductor A′ B′ is continuous; this conductor is electrified by induction.

If the armature A is sufficiently highly charged, the points A′ will discharge + electricity upon the glass, and the points B′ (—) electricity; because the induced charges are attracted through the glass by the charged armatures.

The same phenomenon will be reproduced at least during the first half-turn of the cylinder, for the portions of glass which have received electricity, move

Fig. 4.

away rapidly and permit the influence of the armature A to be again exercised on the conductor.

At this moment the interior surface of the cylinder can be divided by a plane approximately horizontal into two parts oppositely electrified; the upper one (—) and the lower (+). These electrified bodies contribute to increase the production of electricity by a series of reciprocal reactions. For, the armature B *b* by the induction of the two sides has its base B charged with (+) electricity, and its point *b* with (—) electricity, which discharges itself on the exterior surface of the cylinder; the same influence is exercised on the second armature, at the base A of which, the negative charge increases, while + electricity escapes by the point *a*.

During the next half-rotation, the difference of potential at the ends of the conductor will be increased by the electrification of the second armature, the increase of charge of the first, and by the direct influence which is exercised on the points by the electric charge on the surface of the moving cylinder. If we neglect loss, the electrification ought to increase in a geometrical progression; but the apparatus soon attains its maximum power, and the electrification of different portions of it becomes constant.

The two faces of the cylinder are always + at the lower and — at the upper part; but the plane of separation between the charges of opposite signs is not horizontal.

The + flow which escapes from the point A′ encroaches on the negative portion because of the attraction which that portion exercises, and the repulsion of the lower portion.

One sees this by the form of the luminous glow.

If we neglect the difference of properties between the two kinds of electrification, we see that the two portions of opposite kinds, spread on the interior surface of the cylinder, will be separated by a plane *a β*, not parallel to the line joining the points. In the same way the electricity which escapes from the paper points will distribute itself on the exterior surface of the cylinder in two zones separated by another plane *a′ β′*.

stream of sparks passes as long as the motion is continued. One pole gives (+) electricity, and the other (−).

Fig. 5. This engraving represents a large Holtz machine made by Mr. Ladd, having 24 ebonite plates—12 fixed, 12 revolving, each 2 feet in diameter.* It is enclosed in a plate-glass case, the air inside which is kept dry by pumice-stone and sulphuric acid. The handle and the discharging rods are outside, so that the machine can be worked without opening the case. It can

Fig. 5.

be worked either by hand or steam power. It is always ready for use, and gives very fine effects.

THE ELECTROPHORUS. FIG. 6.

Among the various forms of this instrument, we will select for description that made for use with Sir William Thomson's electrometer.

It consists of a circular plate of ebonite on a metal base, from which a brass pin comes through the ebonite and ends flush with

---

* Increasing the number of plates does not much increase the length of the spark but greatly increases the quantity of electricity produced.

its upper surface. A brass plate of the same size as the ebonite has a glass handle fixed to its centre, by means of which it can be laid flat upon the ebonite. The ebonite is negatively charged by friction either with a catskin or silk handkerchief. The brass plate, being laid upon it, becomes charged by induction—the under side positively, the upper negatively. The pin allows the (−) electricity to escape to earth, the (+) being held by the attraction of the negatively charged ebonite; and on removing the brass it will be found to be positively charged. An electrophorus $2\frac{1}{2}$ inches diameter will give about $\frac{3}{16}$ in. spark.

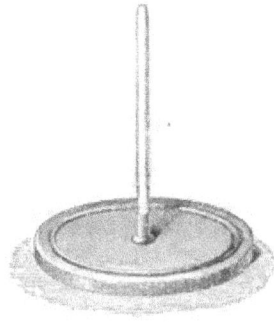

Fig. 6.

A fresh charge can be obtained as often as the brass plate is lifted on and off the ebonite.

When the instrument is made without an earthing pin, the top of the brass has to be touched by the finger each time, immediately before being lifted, in order to allow the negative electricity to escape.

CHAPTER III.

QUANTITIES OF ELECTRICITY.

WE have stated that we only know **electricity as a property** of electrified bodies. We know that we can by certain **means increase** or diminish the electrification of any body. We have **now to** consider if the effects of **such** increase or diminution are the same as would have been produced if a quantity of a something which we call Electricity had been added **to** the body or taken from it. This is a matter of **the** greatest importance, for if such addition and subtraction were not possible, **we** should have no means of measuring electrostatic effects.

" Electrification," **or, as we shall now call** it, " Electricity," may be regarded **as a** quantity. **That is,** for instance :—If **there be two bodies** containing **equal quantities of** one kind of electricity, and the whole of the electricity **in** one be transferred (by any means) to the other, then the latter will contain twice the quantity of electricity which it contained before ; and similarly, if the electricity in **a** charged body be equally divided between it **and one not** previously electrified, **the** resulting charge of each will **be one-half of that** previously **in the** charged body.

For purposes of calculation, electricity of either kind may be treated precisely as if it was a material incompressible fluid.

It may be objected **that it** differs from **a** fluid because equal quantities of two kinds neutralize each other, and that we cannot conceive two material fluids disappearing when **mixed.** We do not **know, however, that the** fluids have disappeared, but **only** that **after mixture** they are no longer appreciable by **our senses.** Suppose a **tank of** tepid water, **and** that hot and cold water are supplied to **it by two taps; and that the** only means of examining the flow is **by** noting the changes of temperature, i. e. that **we** can **only observe one** of **the** properties of the water, and **not all** of them. Then the thermometer will show us if **either the hot** or **the** cold tap is open alone ; but **if both** are open **the** thermometer will give the same indication **as if** both are shut, and we may say that the two fluids have neutralized each other *; and this* will be true with regard to the only **one** of their properties which we had the means of examining.

We must not, **however, commit** ourselves to the idea that electricity is a **substance. We** do not know whether it is or is not. There **are many other** instances of quantities which are not substances.

No one, for instance, supposes pressure to be a substance, and yet nothing is easier than to add two pressures together. Two equal weights in a scale-pan each produce their own pressure, and, when put in it together, produce a pressure double that produced by either of them. Two horses can move a cart too heavy for one, because the pressures exerted by them are added together.

Again, velocity is not a substance, but it is a quantity.

For purposes of calculation, any increase or decrease of electrification may be considered to be produced by the addition or by the taking away of a quantity of something (which we call electricity). This quantity can be added or subtracted by the ordinary rules of algebra, and for the two kinds of electrification has respectively a + or − sign.

To perform any experiments on the addition and subtraction of quantities of electricity, it is necessary that we should have some means of transferring the whole of the charge of electricity in a body, from it to another body. This cannot be done by merely placing the charged and uncharged bodies in contact, for in that case the charge only divides between the two in a proportion depending on their relative sizes and shapes.

HOLLOW CONDUCTOR.

It has been found by experiment that—

If a hollow closed conducting body be charged, however highly, with electricity, the whole of the charge is found upon the outside surface, and none whatever on the inside.

Experiment.—Take a hollow metal globe, with a small opening at the top, or, for rough experiments, an ordinary tea canister. Let it be insulated by being hung up by white* silk threads. Charge it as highly as possible by means of the electric machine.

Hang up a small metal ball by a silk thread, and touch the outside of the canister with it. On removing the ball and holding it near the electroscope, it will be found to be charged in the same way as the canister.

Now lower the charged ball down *inside* the canister. Let it touch the inside, and then be drawn out without touching the edge.

* Coloured threads are sometimes prepared with metallic dyes, which spoil their insulating power.

The ball will be found to be completely discharged.

However highly we charge the hollow conductor, we shall still find that there is no indication whatever of a charge on its inside.*

We **now** see how we **can add together** any number of charges.

Discharge the hollow **body.** Let there be any number of small conductors of any shape, and charged with **any** quantities of **+ or − electricity. Lower** them by **insulating** threads into the hollow conductor, either one at **a** time, or several, or all together. Let them touch the inside, and draw **them** out without letting them touch the neck, and they will **be** found **to be** completely discharged, **and** all the electricity that **was in them** will be found in one **charge** on the outer surface **of the canister.**

Experiment.—Charge **the** suspended ball with + electricity, by holding **it against** the conductor of the electric machine. **Lower** it into **the** canister and let it touch the inside, the canister will now be feebly charged. On each repetition **of** the process, it will be found that the canister is more strongly charged.

This is **because** the successive charges of the ball are *added together* on the canister.

On introducing a small **(−) charge, the** electrification of the canister will be observed to **diminish.**

TOTAL QUANTITY OF ELECTRICITY PRODUCED BY INDUCTION.

It can be proved mathematically that :—

The total quantity of electricity, of the opposite kind to its own, which a **charged** *body induces on neighbouring bodies is exactly equal to its own charge.*

The following experiment illustrates this law :—

Experiment.—Connect the hollow conductor to the electroscope, and, inside it, suspend a positively charged metal ball, but do not allow it to touch the canister. Negative electricity will be induced on the inside of the canister, and positive on the outside.

The latter will cause a divergence of the gold leaves.

Let the outside be now connected to earth for a moment by means of the finger or other conductor.

---

* This will not **be** found to **be strictly true, close to** the mouth of the canister. The mathematical **theory** requires a *closed* hollow body. In experimenting we are obliged to leave an opening for the introduction of the ball.

The leaves will collapse, showing that the positive electricity has been removed.

The negative charge is temporarily held on the inside of the canister by the attraction of the positively charged metal ball, and so produces no divergence of the leaves which are connected to the outside. Now let the ball be removed. The negative electricity is set free, and at once goes to the outside of the canister, and the leaves diverge.

We now have the ball charged with a certain quantity of positive electricity, and the canister charged with the total quantity of negative electricity, which that charge of positive can induce.

To show that these two charges are equal, we must add them together, and the result will be zero.

This may be done in two ways. Either we may allow the ball to touch the canister, or we may lower ball and canister side by side into a larger canister, and let them touch it.

In either case all signs of electrification will disappear.

EQUAL QUANTITIES OF + AND − ALWAYS PRODUCED.

We stated in Chapter I. that when electricity is generated by friction, equal quantities of + and − are always produced.

The following experiment illustrates this fact :—

Experiment.—Place a large friction electric machine (fig. 2) on an insulating stool (i. e. a stool with glass legs), and connect the rubbers to the conductor by a wire. However rapidly the machine is worked, it will not even cause a divergence of the leaves of an electroscope attached to the conductor.*

This is because the negative electricity produced on the rubbers exactly neutralizes the positive produced on the glass.

We remember also that in Chapter I. we showed that the induced charge always consists of equal quantities of the two electricities.

We can now state an important law :—

If any system of bodies, however large, not previously electrified, be insulated, then, whatever electrification be, by any means, developed in any of them, still, provided that no electricity enters or leaves the system, the algebraic sum of the electrifications will remain zero.

---

* The handle should be insulated from the hand by a sheet of indiarubber.

CHAPTER IV.

ELECTRIC FORCE.

LAW OF ELECTRIC FORCE.

A VERY rough experiment will show us that if electrified bodies are some way apart, they will not attract or repel each other so strongly as if they are close together. In other words, when we increase the distance we diminish the force. We are now going to consider at what rate the force diminishes as the distance increases—whether it diminishes at the same rate as that at which the distance increases, or at a faster or a slower rate.

Coulomb investigated this question experimentally by means of the *torsion balance** which he invented for the purpose, and he proved the following law :—

The force of attraction or repulsion between two electrified bodies, whose sizes are very small in comparison with their distance apart, varies inversely as the square of their distance apart ; that is to say, if the repulsion at a certain distance were equal to unity, then at double the distance the repulsion would be one quarter, at three times the distance it would be one ninth, at four times one sixteenth, and so on.

INDIRECT PROOF.

It can also be proved mathematically that this law is the only one consistent with the fact that there is no electricity inside a hollow conductor.

Now as our means of detecting electrification are far more delicate than those for measuring electric forces, the fact of the non-electrification of the inside of a conductor gives a more accurate proof of the law of force than any direct measurement can do.

* See p. 33.

Cavendish,[*] in 1771-81, showed by this means that the law of the inverse squares must be accurate to within $\frac{1}{50}$th part, and our modern electrometers show that it is accurate to within 1 part in 72,000.

FORCE DEPENDS ON CHARGES.

The force between two electrified bodies depends on the charges of electricity on each, as well as on their distance apart. If the one body is charged with a quantity of electricity called e units of electricity, and the other body with one unit, then, when the bodies are at any fixed distance, say one centimetre, from each other, the force between them is found to be proportional to e, the charge on the first body.

It is also found that, if the second body be charged with two units of electricity, the force will be proportional to $2e$ (twice e) ; if with three units, $3e$, etc. Now, suppose it is charged with any number of units, and let us call that number e', the force will be proportional to $e\,e'$ (e multiplied by e'), *or the force between two small electrified bodies one centimetre apart is proportional to the product of their charges.*

The units are so chosen that the force is actually equal to this product, when the distance is one centimetre.

And by Coulomb's law it will at any other distance *be equal to the product of the charges divided by the square of the distance.*

If we write this symbolically, we say that the repulsive force between two small spheres separated by a distance r (which is great when compared with the radii of the spheres),[†] and charged respectively with e and e' units of electricity, is equal to $\dfrac{e\,e'}{r^2}$.

We see that, when the charges are of similar sign, the expression for the force will be $(+)$; and when they are of different signs there will be a $(-)$ repulsion—i.e., an attraction. This expresses the fact that similar electricities repel, and unlike attract.

PHYSICAL NATURE OF ELECTRIC FORCE.

The phenomena of electric attraction and repulsion, as well as

[*] *The Electrical Researches of the Hon. Henry Cavendish,* **F.R.S.** Edited by J. Clerk Maxwell, F.R.S. Cambridge, 1879. Art. 217.

[†] When this is not the case, the inductive action of the two spheres on each other so disturbs the distribution of the electricity on them as to alter the attraction.

those of induction, show us that the influence of a body charged with electricity extends for a considerable distance all round it, for it can act on bodies at a distance from it.

The question, " How is this action conveyed across the intermediate space?" is the most important in all electrical science. If we could answer it completely we should probably be able to comprehend the physical nature of electricity itself.

To say that the effects are due to a "direct action at a distance" is really only re-stating the question.

Although the hypothesis of "direct action at a distance" is a convenient one for mathematical calculation, it conveys absolutely no physical meaning to our minds at all.

We are at once met with the further question, " What is the nature of this direct action?"   " How is it transmitted?"

We at once endeavour to imagine of what the connecting link between the attracting bodies is composed. The hypothesis of direct action at a distance assumes that it is not composed of matter, and we have at least a difficulty in conceiving of a link, strong enough to move material bodies, composed of anything else.

A Medium necessary.

There is no doubt that there must be some physical and material connecting link whenever electrical action of any kind is transmitted from one body to another.

The only manner in which we can in any way account for the observed facts of attraction, repulsion, and induction, is by assuming that *the forces are transmitted by a strain* or distortion of the medium which fills the space between the electrified bodies.*

We have seen that electric forces act through air, glass, and all other insulators, and we have every reason to believe that the forces are transmitted through them by a strain or distortion of their particles.

---

* The word " strain " is in physics defined to mean " an alteration of size or shape." Any alteration of size or shape whatever is called a strain. The word includes compression and extension—as, for instance, that of a piece of stretched india-rubber where the length increases and the breadth diminishes. It includes all alterations of volume, as compression or expansion of a gas, all twisting and bendings, and all vibratory motions other than those of a rigid body as a whole.

It does *not* include the force producing the alteration of size or shape. A force which produces a strain is called a " stress."

Numerous arguments in favour of this hypothesis will be given later on in this book.

When a weight has to be moved by pushing it with a pole or pulling it by a cord, the mechanical action is transmitted in a very similar way, for the portion of the pole or rope nearest the hand is strained by being compressed or stretched. Then, as it recovers its shape, the strain is transmitted to the next portion, and passes on until it and the force reach the body.

The transmission of strain may be very beautifully seen at any railway station where shunting is going on, if a train of carriages is being pushed by an engine which happens, instead of giving a steady pressure, to strike a slight blow on the carriage nearest to it.

The furthest carriage does not at once move, but the buffer-springs of the first are compressed—that is, the first carriage is for an instant "strained" by having its total length, from buffer to buffer, shortened by some inches.

It instantly recovers from this strain, by the expansion of the springs; but as it cannot expand towards the engine, it expands away from it, and transmits the strain to the next carriage by compressing its buffer-springs, and the process is repeated till the force is transmitted all the way from the engine to the carriage furthest from it.

In this case of transmission of mechanical force, the strain is a simple compression and expansion, but in the case of the electrical forces we have no means of knowing its exact nature; it is probably a much more complicated phenomenon.

THE ETHER.

In addition to being transmitted through air, glass, and other insulators, it is found that the electric forces are transmitted, not only across the best vacua we have as yet been able to produce artificially, but certainly also across the inter-planetary spaces.

There is no doubt that the earth is affected by electrical phenomena occurring in the sun.[*]

Now we know that in these spaces there is no matter such as we are commonly cognizant of, and we must therefore suppose

[*] See Chapter XVI.

them to be filled with matter in an excessively attenuated state.

We call this matter " *Ether*," and suppose it to be a fluid many million times thinner than air, and having very great elasticity.

In the fourth part of this work we shall endeavour to show that this ether is the same medium which conveys the light and heat from the sun to the earth—that is, that Light, Radiant Heat, Electric and Magnetic Induction are all different disturbances of the same ether-sea.

We must consider that this medium transmits electric forces, but does not in general exhibit electrical properties of its own.

We must consider the ether all round a charged body to be in a strained state; but we know that no electrical properties are exhibited at any point near the charged body, until a portion of ordinary matter is placed at that point, and then the matter receives electrical properties from the medium immediately surrounding it.

The strength of these electrical properties depends on the amount of strain of the medium at the point.

We may, if we choose, consider that the ether permeates all bodies.

We shall presently show that electric forces are transmitted through different insulators with different strengths.

We may either consider that electric forces are transmitted through ordinary bodies by strains of their particles, and only through so-called empty space by strains of the ether; or we may say that all electric forces are transmitted by strains of the ether, but that the ether in different insulators is modified in some way which will account for the difference of transmission.

As we have as yet no experimental means of comparing the merits of these two hypotheses, we may adopt whichever seems most convenient for purposes of calculation. On the whole I think the latter is the most useful.

POINTED CONDUCTOR.

Experiment—Fix a fine metal point to the conductor of the electric machine and work the machine.

It will be impossible to collect any appreciable charge on the conductor, the electricity all escapes by the point.

EXPLANATION.[*]

Let M P N, fig. 7, be a section of a pointed charged conductor, and let us consider the forces tending respectively to drive off a unit of electricity from a position S on the side of the conductor, and from its point P.

The electricity on the side M P has no effect in tending to cause an escape of electricity from S, for no part of it can act on S in a direction perpendicular to the surface, M P. Such portion of the electricity on the side N P as is moderately near to S will act on it. In fig. 7 we see that there is no electrified

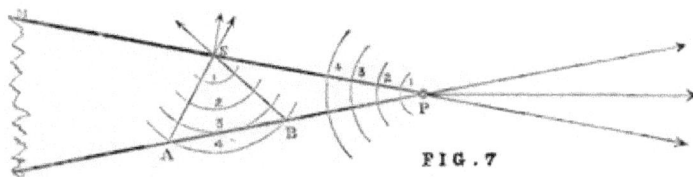

FIG. 7

surface within a distance of 1 or 2 of S, a very small portion within a distance 3, and a portion A B within a distance 4. We see that, as the force diminishes as the square of the distance. there is only a very small force acting on S, and even that acts at such an angle to the surface M P that only a small fraction of it acts perpendicular to M P; and further, when, instead of a section of the pointed conductor, we consider the whole of it, we shall find that, as we get nearer to S, the electricity always acts more diagonally upon it.

Upon P, on the contrary, there is electricity acting at all the distances from zero to 4 and further; and it all acts almost directly in the direction tending to drive off electricity from the point P; and this is still the case when we consider the whole conductor instead of the section.

There will therefore be a very much greater force tending to drive electricity from a point than from any other portion of a conductor.

LIGHTNING CONDUCTORS.

The same reasoning holds if the air is the charged body and the metal the way of escape, for in the above experiment we

---

[*] The following proof is due to Mr. W. E. Ayrton.

have no means of knowing whether the diminution of **the electrification of** the pointed conductor **is** due to a loss of **one kind of** electrification or to a gain of the **other** kind.

This is the reason why lightning conductors are made pointed. **The force** at their ends is very **great, and** therefore the **electricity** of the **air is** quietly carried away to the ground. **At** first most conductors **were made with** knobs. **Then** the electricity accumulated at the surface until the electric **force was** sufficiently **great** to cause a spark. It was held that **the knobs must be the most** efficacious, because the lightning **was seen to** strike them, and never struck the **points.** The fact that a point prevents the lightning from **ever** striking **at all was** not known.

CHAPTER V.

DENSITY.

THE more intense the electrification of a body, the greater the electric force, and the greater the tendency of the electricity to escape to surrounding objects.

The intensity of the electrification at any point of a body or a surface is called the *Electric density* at that point.

The charge of a body is the quantity of electricity in it.

The volume density at any point of a body, when the density is uniform, is the charge or quantity of electricity in each cubic centimetre of volume.

When the density is not uniform, the volume density at any point is equal to the quantity of electricity which there would have been in the cubic centimetre surrounding the point, if the density at every point of that cubic centimetre had been the same as that at the given point.

The surface density at any point of a surface, when the density is uniform, is the quantity of electricity on each square centimetre of surface.

When the density is not uniform, the surface density at any point is equal to the quantity of electricity which there would have been in the square centimetre surrounding the point if the density at every point of that square centimetre had been the same as that at the given point.

The force with which the electricity endeavours to escape from any particular portion of a surface increases with the density at that portion of the surface.

CHAPTER VI.

POTENTIAL.*

DEFINITIONS.—Whenever electricity moves, or tends to move, from one place to another, there is said to be a " *Difference of Potential* " between those two places.

The place *from* which positive electricity tends to move is said to be of higher potential than the other.

Suppose a quantity of electricity to flow from the one point to the other, then *the " Difference of Potential," or as it is also called the " Electromotive Force," between those points is a quantity which expresses the amount of work which each unit of the electricity could do on its journey, if it were all utilized by being passed through some perfect machine of which it formed the motive power.*

Difference of potential is calculated as follows :—

Let a unit of electricity be forced from the one place to the other, in the opposite direction to that in which the electric forces tend to move it ; " *Mechanical Work* " will have to be done on it by a man, a steam-engine, or other source of power.

The difference of potential between any two points is defined to be numerically equal to the amount of work required to force a unit of electricity from the one point to the other, in the direction opposed to that in which it tends to move.

The expression " Potential at a Point " is an abbreviation for " Difference of Potential between the point and the Earth." The potential of the Earth is taken as the standard, and is called zero.

In the same way the expression " height of a roof" is an abbreviation for " difference of height between the roof and the

* The definitions and statements given in this chapter apply to statical electricity. In their application to voltaic electricity, certain limitations have to be introduced which will be noted in their proper place.

surface of the earth." The height of the earth's surface is taken as the standard and called zero.*

Very low potentials, that is potentials much lower than that of the earth, are sometimes called high negative potentials.

We know that similarly charged bodies repel each other—that is, that electricity has a tendency to move from points near a charged body of similar sign to points further off.

We therefore say that points near a charged body are of *numerically* higher potential than points further off.

When we know the charge of the body, the distance of each of any two points from it, and the law of force, we can calculate the difference of potential between those points by means of the Integral Calculus.†

The result of the calculation is, that if e be the charge of the inducing body, r_1 and r_2 the respective distances of the points from its centre, we shall have

$$\text{Difference of Potential} = \frac{e}{r_1} - \frac{e}{r_2}.$$

If one of the points r_2 is on the earth, then $\frac{e}{r_2}$ is the potential of the earth, and the " Difference of Potential " between the other (r_1) and the earth is $\frac{e}{r_1}$, and we have agreed to call this the " Potential at r_1," and we write it V_{r_1}.

It can be shown mathematically that *the potential at any point*

---

* A roof 100 feet above the ground is thus said to be 100 feet high, but if we had taken the centre of the earth for our standard, we should with equal correctness have said that the roof was about 4000 miles high.

† Let V be the potential, e the charge, r_1, r_2 the distances respectively, we have (remembering that the attraction $= -\frac{e}{r^2}$),

$$V_{r_1} - V_{r_2} = \int_{r_2}^{r_1} \left(-\frac{e}{r^2}\right) dr = \frac{e}{r_1} - \frac{e}{r_2}.$$

When one of the points is on the earth, we assume

$$\frac{e}{r_2} = 0,$$

or that the potential of the earth is zero, which gives

$$V_{r_1} = \frac{e}{r_1}.$$

due to two or more charged bodies is the algebraic sum of the potentials due to each.*

If two conductors, which are of different potential, be connected by a wire or other conductor, electricity will flow from one to the other as long as the potentials continue to differ.†

If, *by the expenditure of work,* the difference of potentials is kept up, the flow may continue indefinitely; but if no work is expended, all parts of the conductor instantly reach the same potential; and we may state that—

All portions of a conductor of any one material, subject only to statical forces, are always at the same potential.

For, by definition, a conductor is a body in which electricity can obey any tendency to move, and, in a conductor, electricity goes on moving from places of high to places of low potential, until there is no longer any tendency to move—that is, until all the conductor is at the same potential; and this adjustment takes place in an extremely small fraction of a second.

Electro-Statics and Electro-Kinetics.

As long as a flow of electricity continues, and is kept up by the expenditure of work, the conductor has properties different from those of a simple electrified body. That branch of electrical science which treats of the properties of simple electrified bodies is called *Electro-Statics,* because in them the electricity is supposed to be at rest; and that branch which treats of the flow of electricity is called *Electro-Kinetics.* See Part III.

Equipotential Points, Lines, and Surfaces.

Points whose potentials are equal are called "Equipotential Points." If a series of such points form a line, it is called an "Equipotential Line." If the potential at every point on a surface is equal, it is called an Equipotential Surface.

Equipotential Surfaces.

The potential due to a charged body has its greatest numerical value at the body itself, and diminishes in all directions.

---

* For the potential is calculated from the force acting, and when two or more forces act at once, their effect is the sum of the effects due to each.

† The two conductors being of the same metal and at the same temperature. (See Chapter XXXIV.)

Let us take some given point near any one body where the potential has a certain value. There will, in general, be a number of other points where it has the same value. If, for instance, our first point is to the right of the body, there will probably be a point of the same potential at a certain distance on the left hand of it, and others above and below, &c.

In general there will be an infinite number of points of the same potential situated all round the charged body.

Let us imagine a surface to be drawn through all the points whose potential has some particular value. It will form a closed surface all round the body.

Such a surface or shell is called an " Equipotential Surface," because at all points of it the potential is equal.

Now, if we construct several such surfaces for different values of the potential, we shall have a series of shells, each enclosing the one corresponding to the next higher numerical value of the potential.

When there is only one electrified body, and that a sphere uniformly electrified, the equipotential surfaces become a series of hollow spheres, whose common centre is the centre of the electrified body.

When there is more than one electrified body, or when it is not spherical, the surfaces are of less simple form.

There is no force tending to move electricity from one portion of any equipotential surface to any other portion of the same surface; and a charged body can be moved from any one point on it to any other without the expenditure of any work at all.

For whenever there is a tendency to move, there is, by definition, a difference of potential; and whenever work has to be expended to move a charged body, there is, by definition, a difference by potential. In the case of an equipotential surface there is no difference of potential, and, therefore, no tendency to move and no work expended.

The surface of every conductor of any one material, acted on only by statical forces, forms an equipotential surface.*

LINE OF FORCE.

Definition.—The line of Force at any point is the direction in which a charged body placed at that point tends to move.

* See page 28.

The line of force at any point is always perpendicular to the equipotential surface passing through that point.

For, if the direction of the force is inclined to the surface, the force may be resolved into two—one perpendicular to the surface, and one along it. But we have just shown that the force along the surface is zero, **and** therefore the whole of **the** force is pependicular to the surface.

Relation between Potential and Force.

It can be proved mathematically that—

The force at any point in any direction is equal to **the rate at** *which the potential begins to change, as we begin to leave that point in the given direction.* *

Potential and Charge.

The electrifications of different **parts of** the same conductor, when acted on by inductive forces, may be very different, but the **potential of** every part of it will still be the same.

Thus, if a conductor (A) be placed on an insulating stand, near an electric machine **(as on page** 6), electroscopes placed **at** its two ends will diverge with **opposite** charges, while one placed at the **middle** would not diverge at all. But if an electroscope be **placed at a** distance, and a wire **brought** from **it** to the conductor (A), **then the divergence of** the electroscope will depend **only on the potential to which** the conductor (A) has been raised **by the** induction **of the machine**, *and will be exactly the same at whatever point* **of the conductor** (A) *the wire is attached*. Hence an electroscope, protected from direct action,† and attached to any conductor by a wire, **measures** the potential of that conductor.

---

* For rate of change is $\dfrac{d\,V}{d\,r}$. Let f be the force; we have by definition

$$V = \int f\, dr + C, \text{ whence } \frac{d\,V}{d\,r} = f.$$

† See page 5 (Metal **Screens**).

CHAPTER VII.

ELECTROMETERS.

ELECTROMETERS and Electroscopes are instruments for measuring the strengths of the attractions and repulsions between electrified bodies.

By measuring the force of repulsion between two charged bodies at different distances, Coulomb established the law according to which electric force diminishes with the distance. (Page 18.)

By means of a suitable electrometer we can also measure the *potential* of a charged body, or that of any point in its neighbourhood. (See preceding page.)

The two earliest types of electrometers were Cavendish's and Lane's.

CAVENDISH'S ELECTROMETER.*

This was invented between 1771 and 1781, and consists of two pith balls hung, touching each other, by two linen threads. When similarly electrified, they repel each other, and diverge more or less according to the strength of the electrification.

LANE'S ELECTROMETER.†

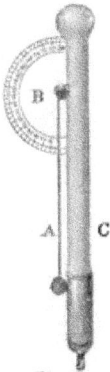

Fig. 8 is copied from the *Phil. Trans.* 1772. A wooden stem C is mounted in a metal socket, which can be screwed into the conductor whose electrification is to be measured. A pith ball, fixed to a straw stem A, hangs from a pivot at the centre of a divided semi-circle B.

Electricity is communicated from the metal socket to the ball,

---

* Cavendish's *Electrical Researches*, edited by J. Clerk Maxwell, F.R.S. Camb. Univ. Press, 1879. Art. 244.

† Ibid., Art. 559.

which is repelled. The number of degrees over which the straw
passes gives a rough idea of the strength of the electrification.

THE GOLD-LEAF ELECTROSCOPE.

The gold-leaf electroscope (fig. 9) consists of two strips of gold
leaf, which in a large instrument may be 4 inches long and 1
inch wide, hung together by their upper ends to a metal rod. This
rod is fixed through a hole in the top of a glass shade, inside

Fig. 9.

which the gold leaves hang. The upper end of the rod terminates
in a flat brass plate.

When an electrified body is placed on the plate, a portion of
the electricity passes to the gold leaves, and charging them both
with the same kind of electricity, causes them to repel each other
and diverge as shown in fig. 9.

Owing to the lightness of gold leaf, this is a sensitive instrument for detecting the presence of small quantities of electricity.

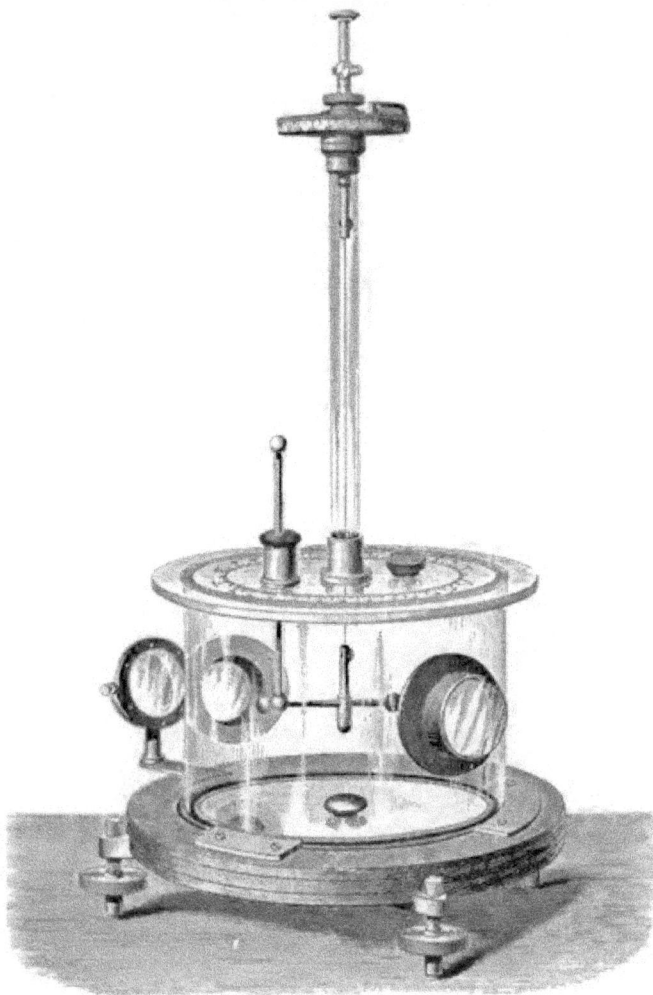

PLATE I.—COULOMB'S TORSION BALANCE.

COULOMB'S TORSION BALANCE.—PLATE I.

The first great improvement in electrometers was made by Coulomb, who, about 1785, published an account of an electrometer which, from the principle of its construction, is called the "*Torsion Balance.*"

The *torsion* exerted by a thread, suspended vertically, is the force tending to twist the lower extremity, when the upper extremity is turned through an angle.

If a weight be suspended by a piece of string, and the upper end of the string twisted between the finger and thumb, the weight will begin to revolve.

If the weight is prevented from revolving by the fingers of the other hand, a certain force will have to be exerted to stop it. This force is equal in amount, but opposite in direction, to the force exerted by the torsion of the thread. The amount of force called into play by turning the top of the thread through a certain angle depends on the thickness and length of the thread and the weight suspended from it. It is obvious that, by making the thread long and thin and the weight small, a very large angle of twist can be made to exercise a very small force.

In Coulomb's torsion apparatus (Plate I.) a very long fine thread is suspended in a vertical tube. At the top of the tube is a horizontal circle divided into degrees called the torsion circle, which can be turned by means of a short vertical rod fixed to it through its centre. A pointer, or in delicate instruments a vernier, fixed to the tube shows the position of the circle. The thread is attached to the vertical rod passing through the circle. Turning the rod twists the top of the thread through an angle which is shown by the vernier.

If a greater angle than 360° is required, the circle must be turned more than once round, and a record of the number of turns kept. At the bottom of the thread is attached a light rod, usually a stout straw, arranged so as to hang in a horizontal position. At one end is a ball of pith, gilt. This gives a spherical conducting surface with very little weight. It is balanced by another ball at the other end of the straw. The vertical tube, carrying the circle, stands on a horizontal glass plate, which has a hole in it to allow the thread to pass through, and forms the lid of the lower part of the case. This lower

D

part is large enough to allow the horizontal arm to swing freely. On the top plate is engraved a circle divided into degrees : the base consists of a piece of looking-glass. At one part of the top plate, at a distance from the centre equal to the radius of the straw arm, is a hole through which **another** gilt pith ball fixed to a vertical stem **can** be put.

The position of the suspended **straw can be observed by** means of the engraved **circle.** The observer, **looking down, moves** his eye until a division of **the circle, the straw, and the reflection of the** same division in the looking-glass, are all **in the same straight line.**

To use the instrument the fixed **pith ball is removed, and the** torsion circle turned till the suspended **pith ball occupies exactly** the position formerly occupied by the fixed one. **When it has** come to rest in this position there is no torsion, and the reading of the torsion circle is taken as the zero. The fixed pith ball is then electrified **and put in position,** pushing the suspended ball to one side, **and at the same** time communicating half its charge **to** it. **The balls now repel each other, and if the length and thickness** of the thread and the strength of the charge have been properly adjusted, the suspended arm should turn through from 30° to 45°.

The position of the straw is noted, and the torsion circle is turned so as to force the balls towards each other, until the straw **pointer has** moved through an **angle equal to one** division of the engraved circle. **The number of degrees through which** the torsion circle has been **turned is then noted, and the process is** repeated for several divisions **until the balls are forced rather near together.** *A table can then be formed showing the force of repulsion correspond-ing to each decrement of distance,* for the force overcome in each case is simply proportional to the number of degrees through which the torsion circle has been turned.

A slight modification of the arrangements **will enable the** force of attraction to be measured when the two **balls are** oppo-sitely electrified.

The Torsion Balance is not at all an easy instrument to use ; it is **not very delicate,** nor is it accurate enough for modern re-quirements. Its great defect is that the whole electrical force in it is that due to the electrification to be measured, which we remember is divided between the fixed and moving balls. When, as is often the case, the **electrification is very** small, the balance shows a corresponding **want** of sensibility.

The problem to be solved for the improvement of electrometers was to obtain a considerable force for a very small quantity of electrification. Now we know that at any given distance the force between two charged bodies is directly as the product of their electrifications, so that, if we double the charge of one and halve that of the other, we should still have the same force.

To practically take advantage of this fact in the torsion balance, we should have to give the sphere attached to the suspended arm a high and constant charge. The force between it and the fixed sphere would then be considerable, even when the charge of the fixed sphere was very small, and would be simply proportional to that charge, whatever it might be.

This arrangement cannot be carried out in Coulomb's form of the torsion balance, but is the basis of Sir W. Thomson's beautiful quadrant electrometers, which we will now describe in some detail.*

SIR WILLIAM THOMSON'S QUADRANT ELECTROMETERS.

The principle on which the quadrant electrometers are constructed is as follows :—

A metal needle, made of aluminium for lightness, and of the shape NN (fig. 10), is suspended so that it can turn in a horizontal plane like a compass needle.

This needle is highly charged and, in order to replace the electricity lost by actual leakage, it is connected to a Leyden jar. We shall describe the Leyden jar later (page 61); but it is sufficient for our present purpose to know that it is a kind of reservoir, in which a large quantity of electricity can be stored up. The needle is suspended by an ar-

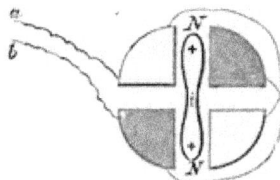

Fig. 10.

rangement analogous to a torsion thread, which tends to bring it back to zero after it has been displaced, and immediately below † the needle are four metal quadrants placed horizontally, as in

---

* *Report on Electrometers and Electro-static Measurements.* Sir Wm. Thomson, F.R.S. *Papers on Electro-statics and Magnetism,* page 260. Macmillan, 1872.

† In fig. 10 the needle is placed between the quadrants for clearness, but its real position is that of fig. 11.

fig. 11. Each is insulated from the one next it, but connected to the one diagonally opposite to it, as shown in fig. 10.

Now, suppose the needle to have a charge of (+) electricity. Let the unshaded quadrants be connected to earth, and the shaded ones, by means of the wire *b*, fig. 10, with the conductor whose electrification is to be measured. Electricity then flows from the charged conductor to the shaded quadrants until they are of the same potential as itself.

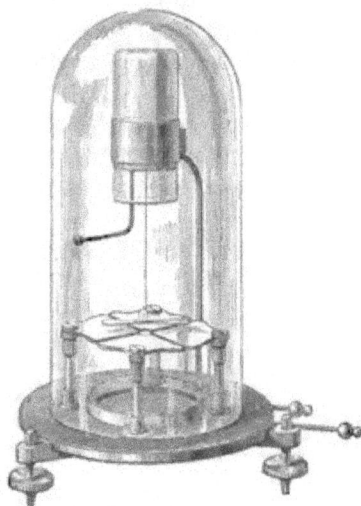

The direction of the deflection at once shows whether the electrification under examination is (+) or (—), for we see that if it is (+) the ends of the needle will be repelled by the shaded quadrants (fig. 10), and the deflection will be to the left, while if it is (—) they will be attracted, and the deflection will be to the right. If the potential is high—that is, if the body under examination be strongly electrified—the deflection will be large; if the potential is low, it will be smaller.

Fig. 11. Lecture Model.

If the charge of the needle be always constant, the deflection will be always the same for the same potential; and if a body has been charged to a given potential one day, it can be brought to the same potential any other day, by attaching it to the electrometer, and varying its charge till the deflection is the same as it was before.

If the shaded quadrants are charged positively and the unshaded negatively, the action of all four quadrants will be to turn the needle to the left, and the deflection would be that due to the numerical sum of the potentials of the two opposite electrifications—that is, to their algebraic difference.

The apparatus can be used for adjusting two potentials to equality; for if two similarly electrified bodies be attached to the shaded and unshaded quadrants respectively, they will tend to turn the needle opposite ways, and the deflection will depend on their difference of potential. If now one of the electrifications be varied till there is no deflection, we shall know that the potentials have been adjusted to exact equality. The electrometer is perhaps more used for this than for any other purpose. ·This method of working is called the " zero method."

PRACTICAL FORMS OF THE ELECTROMETER.

The instrument shown in fig. 11 is only a lecture model. The simplest form that is used for real work is known as the " Elliott-pattern Thomson Electrometer." *

THE ELLIOTT-PATTERN ELECTROMETER.

This instrument is represented in fig. 12 ; it differs from the lecture model in that its metal quadrants are quarters, not of a disc, but of a kind of "pillbox," inside which the needle hangs. Both sides of the needle are thus acted upon.

The Leyden jar is placed at the bottom of the instrument. It contains strong sulphuric acid, and the connection between it and the needle is made by a platinum wire attached to the needle and dipping into the acid. The acid, by its affinity for moisture, keeps the inside of the apparatus always dry. †

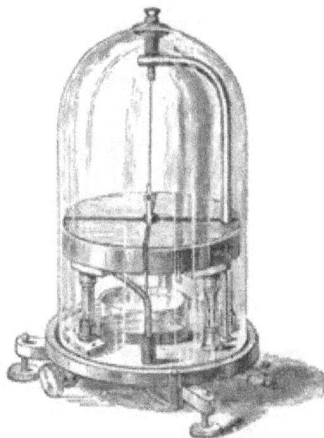

Fig. 12. Elliott-pattern.

Three metal rods project from the instrument. Two (seen on the right) are connected respectively to the two pairs of quadrants, and the third (seen in the front of the figure) can be connected to the needle when it is desired to charge it with electricity.

The needle is suspended by what is called a " bifilar suspension "—that is, it is hung by two fine silk threads, side by side,

---

* It is made by Messrs. Elliott Bros., of London.

† When the instrument is in use the acid must be watched, as it is apt so to increase in quantity by the absorption of moisture as to overflow the jar after a few weeks.

and about $\frac{1}{20}$ inch apart. After the needle has been displaced from its position of rest, these threads always tend to bring it back.

The position of the needle can be adjusted by turning the head at the top of the glass case to which the threads are attached.

The instrument, when in use, is covered by a wire cage connected to earth, to protect the quadrants from the induction of neighbouring charged bodies.*

The Elliott-pattern is extremely sensitive as an electroscope.

It is, I think, the best pattern to use for all that large and important class of investigations where two potentials have to be adjusted to equality by the zero method (page 37); but as there are no means of ascertaining the charge of the needle, or of keeping it constant, it is useless for experiments where potentials have to be measured by observations of the amount of the deflection.

THE LAMP, SCALE, AND MIRROR.

To detect and measure small angular deflections of a needle, a long pointer is necessary; but, if a long material pointer were attached to the needle, its weight would destroy the sensitiveness of the instrument.

Sir Wm. Thomson has therefore arranged a method by which a beam of light is made to act as a pointer of any length, and absolutely without weight.

A circular mirror, about $\frac{1}{3}$ of an inch in diameter, is rigidly attached to the needle by a strip of platinum, which projects up above the quadrants.

Fig. 13.

A lamp and scale, of which the back (that is, the side furthest from the electrometer) is shown in fig. 13, is placed on the table about two feet from the instrument. The light passes through a small opening in the lower part of the scale, falls on the mirror, and is reflected on to the upper part, making a spot of light. The least motion of the needle and mirror, of course, moves the spot along the scale. The distance which it moves is equal to that which would have been

* Page 5.

REPLENISHER

INDUCTION PLATE

PLATE II.—THOMSON'S QUADRANT ELECTROMETER.

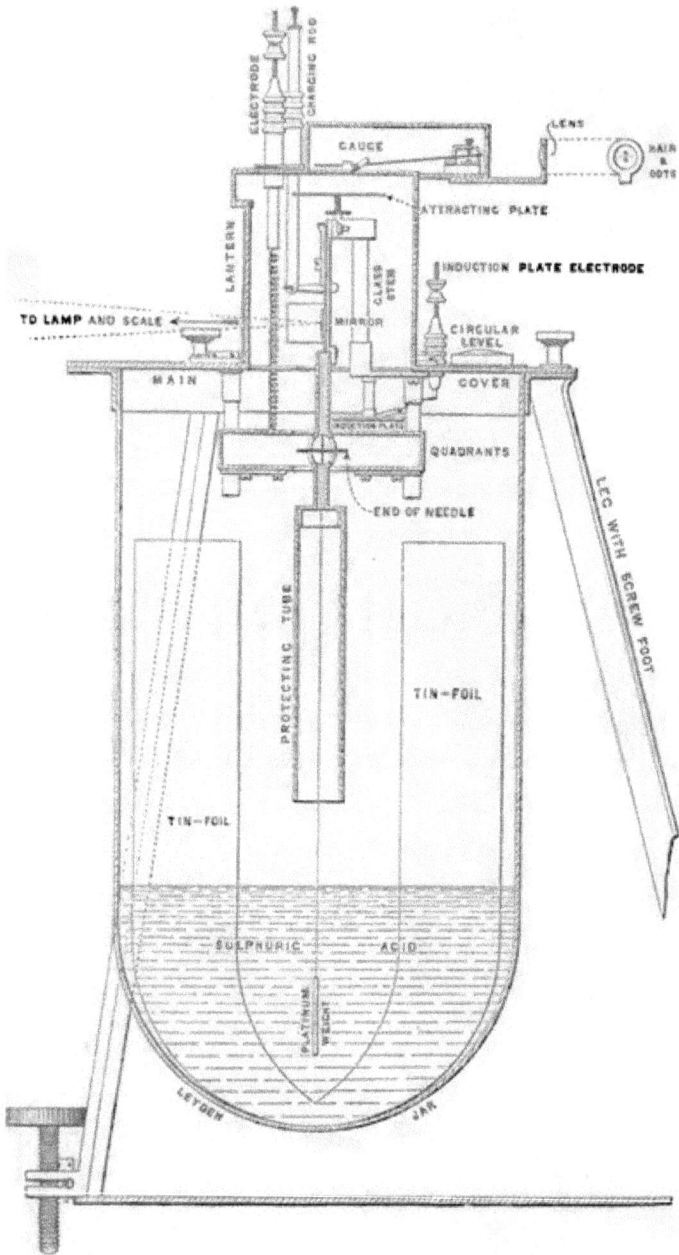

PLATE III.—SECTION OF THOMSON'S QUADRANT ELECTROMETER.

traversed by the end of a pointer whose radius was double the distance from the mirror to the scale.

The aperture through which the light passes is sometimes a vertical slit, sometimes a round hole, with or without a vertical wire stretched across it.

Sometimes the mirror is plane, and the light is brought to a focus on the scale by means of a lens. Sometimes the mirror is concave, and the lens is dispensed with.

When the slit is used, the moving image is a vertical line of light; when the hole is used, it is a bright disc crossed by a fine vertical black line, the image of the wire.

The scale is usually divided into millims., and printed black on white glazed paper.

In using a flat-wicked paraffin-lamp, the wick should be placed " edgeways"—that is, at right angles to the scale.

WHITE'S PATTERN.—PLATES II. AND III.

This is the most elaborate form of the Thomson quadrant electrometer, and is made by White of Glasgow.

The chief requirements of an instrument designed for actual measurement of potentials, and observations of deflection, are :—

(1) A means of testing the constancy of the electrification of the suspended needle, and a method of adjusting this electrification very accurately—that is, of increasing or diminishing it by a very small amount.

(2) A method of varying the directive force tending to make the needle return to its position of rest, and a certainty that, to whatever value this may be adjusted, it will remain constant.

(3) A very accurate method of measuring the deflection of the needle.

We shall shortly see how all these requirements are satisfied.

Plate II. shows the apparatus in perspective, Plate III. in section.

The body of the instrument consists of an inverted glass shade, supported on three legs furnished with levelling screws. The outside of this is coated with tinfoil, and the inside is partly filled with strong, pure sulphuric acid. This, being a conductor, forms the inner coating necessary to make the glass shade act as a

Leyden jar.* Openings are left in the outer coating of tinfoil, where necessary, to allow the interior works to be seen.

The use of the sulphuric acid is to dry the working parts inside the case, as in the Elliott pattern. On the top of the glass is a

Fig. 14.

flat metal cover, which we shall in future call the "main cover." On the main cover is a lantern-shaped erection of brass with a flat glass front. In future we will call this the "lantern."

From near the top of the lantern is suspended the needle, which hangs below the main cover. The quadrants form a box, and completely enclose the needle, as in the Elliott pattern, but are much smaller. In order to connect the needle with the inner coating of the jar, a platinum wire hangs down from it to the acid. To the wire is attached a platinum weight which hangs below the surface of the acid. This, by its friction in the acid, also helps to check the oscillations of the needle. The wire is protected by a wide metal tube.

Three rods project above the main cover; two, called the "Electrodes," are connected to the two pairs of quadrants respectively, and the third is used for charging the needle.

The electrodes can be disconnected from the quadrants by raising them. The left-hand one in Plate II. is shown raised.

One of the quadrants can be adjusted by means of a micrometer screw, whose head is seen in Plate II., and at the left side of fig. 14.

        * See page 61.

Fig. 14 is a drawing of the " main cover," as seen from below. In the centre are the quadrants, and the tube inside which passes the wire from the needle to the acid. The replenisher (*r*) (see page 42) is on the left, and the "Induction plate (*i*)" (see page 44) is above one of the quadrants on the right.

THE GAUGE.—Fig. 15.

The arrangement for knowing whether the potential of the jar and needle remains constant is as follows :—

Near the top of the lantern is a fixed horizontal metal plate connected with the needle and the inside of the jar, but insulated from the rest of the instrument (see Plate III.). Above this is an arm turning on a horizontal axis. The lighter end consists of a square piece of sheet aluminium, which, when the arm is horizontal, lies parallel to the fixed plate and just above it. It is charged by induction oppositely to the plate, and is therefore attracted by it.

Its horizontal axis consists of a platinum wire passed through two holes in the plate and stretched over a little ridge on it. The heavier end of the lever terminates in a horizontal fork with a hair* stretched across it. A piece of white enamel, fixed to the stand, and having two dots on it, projects up through the fork. The hair and the dots are looked at together by a lens.

Fig. 15.

The balance is so adjusted that, when the proper charge has been communicated to the needle, the hair appears exactly between the dots. If now the charge increases, the plate is attracted with greater force, and the hair is seen on or above the upper dot. If the charge diminishes, it descends to the lower dot.

* It is found that a hair from a black and tan terrier gives the best definition.

A precisely similar gauge is used in the absolute electrometer, page 55. Fig. 15 has been drawn from the latter instrument.

Thus, any change in the potential of the needle can be always observed.

THE REPLENISHER.

The replenisher is used to alter the charge of the needle if

REPLENISHER
Fig. 16.

necessary, and works by transferring small portions of the charge to or from the needle and jar as may be required.

By means of it the hair is adjusted to lie exactly between the dots.

The replenisher is shown in fig. 16. Two fixed metal plates, connected respectively to the needle and the earth, act inductively on metal pieces carried on a revolving shaft. The charge induced on the latter is taken off by springs, which touch them at the proper point in the rotation, and conveyed to or from the needle according to the direction of rotation.*

A few moments' use of the replenisher will supply the loss of 24 hours.

A precisely similar replenisher is also used in the absolute electrometer, page 55. Fig. 16 has been drawn from the latter instrument. The two plates at the bottom of the picture belong to it, and not to the quadrant.

THE SUSPENSION.

In the early instruments the needle was suspended by a single silk fibre, and brought back to its position of rest by means

---

* Two other springs put the two carriers into connection once every half turn.

of a small magnet attached to it and a large steel magnet outside the case.

In the modern instruments the bifilar suspension is always used.

By means of an ingenious arrangement of screws, the distance apart of the threads (on which the sensitiveness of the needle depends) can be varied from about $\frac{1}{20}$ to $\frac{1}{8}$ inch. The suspending threads consist of two fibres of unspun silk. Fig. 17 gives the details of the suspension.

The stiff platinum wire which carries the mirror and needle has a cross-piece at its upper end, to which are attached the lower ends of the two suspending silk fibres; the other ends being wound upon the two pins c, d, which may be turned in their sockets by a square-pointed key, to equalize the tensions of the fibres, and make the needle hang midway between the upper and under surfaces of the quadrants. The pins c, d, are pivoted in blocks carried by springs e, f, to allow them to be shifted horizontally when adjusting the position of the points of suspension. The screws a, b, which traverse these blocks, have

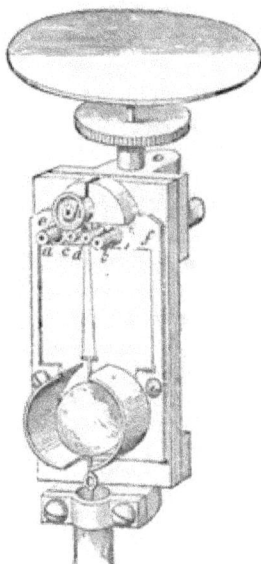

Fig. 17.

their points bearing against the fixed plate behind, so that when a or b is turned in the direction of the hands of a watch, the neighbouring point of suspension is brought forward, and conversely. The needle may thus be made to turn through an angle, till it lies in the symmetrical position represented in fig. 10, when all electrical disturbance has been guarded against by connecting the quadrants with the inside and outside of the jar. The conical pin h passes between the two springs and screws into the plate behind; by screwing it inwards the points of suspension are made to recede from each other laterally, and the sensibility of the needle to a deflecting couple is diminished, and conversely. At the top of fig. 17 is seen the plate which attracts the gauge.

THE INDUCTION PLATE.

The instrument is extremely sensitive, and is primarily intended for the measurement of very small differences of potential. In order that somewhat larger differences may be measured, the *Induction Plate* is introduced to diminish the sensitiveness.

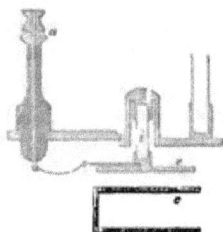

Fig. 18.

Fig. 18 represents a vertical section through it; *c* is one of the quadrants, *e* the induction plate, *i* the glass stem which supports it ; *a* is the electrode.

When it is desired to measure an electrification too high for the ordinary arrangement of the instrument, the electrified body is connected to the induction plate instead of to the quadrant; and so the potential of the quadrant, instead of being that of the body, is only that induced by the charged plate, which is small and at a considerable distance from the quadrant.* See also Plate III. and *i* fig. 14.

* Sir William gives the following directions for further grades of diminished power :—
The connections may be further varied so as to produce other degrees of sensibility giving indications perfectly trustworthy and available for comparative measurements. The different methods of forming the connections, with or without an inductor, are indicated in the following table, where R means the electrode of the pair of quadrants marked RR' in fig. 19, L that of the pair LL', and I that of the induction-plate; C is the conductor led from one of the bodies experimented upon, O the conductor led from the other and connected to the outer metallic'case of the instrument, which may be insulated from the table if necessary by placing a small block or cake of clean paraffin under each of the three feet on which the instrument stands; (R) or (L) means that the electrode of RR' or LL' is to be raised so as to be disconnected from its pair of quadrants. Thus in the grade of diminished power or sensibility standing first in the table on the right, the electrode L is raised, one conductor is connected with R ; I and the other with the case of the instrument. The grade standing last in the table, in which L and R are both raised, is the least sensitive of all. In each of these methods the correctness of the indications has been verified by measurements taken simultaneously with the Standard Electrometer [a modification of the absolute electrometer described on pages 55—58], the measured difference of poten-

THE SCALE.

The deflections are measured by a lamp and scale similar to that used with the simpler instrument, except that the scale, intials being that of the earth and of a Leyden jar fitted with a replenisher, by means of which its potential was varied so as to make the deflected image stand at all points between the extremity of the scale and the zero position. The working of the replenisher being suspended at intervals to allow an accurate reading to be taken of the position of the image and the indication of the Standard Electrometer, the subsistence of a correct proportion between the deflection and the measurement obtained from the Standard Electrometer was verified at all points of the range.

WITHOUT INDUCTION PLATE.

Full Power.

$$\begin{bmatrix} LC \\ RO \end{bmatrix} \text{ or } \begin{bmatrix} RC \\ LO \end{bmatrix}$$

Diminished Power.

$$(L)\begin{bmatrix} RC \\ O \end{bmatrix} \text{ or } (R)\begin{bmatrix} LC \\ O \end{bmatrix}$$

WITH INDUCTION PLATE.

Full Power.

$$\begin{bmatrix} LC \\ RO \end{bmatrix} \text{ or } \begin{bmatrix} RC \\ LO \end{bmatrix}$$

Grades of Diminished Power.

$$(L)\begin{cases} \begin{bmatrix} RC \\ IO \end{bmatrix} \\ \begin{bmatrix} RIC \\ O \end{bmatrix} \\ \begin{bmatrix} IC \\ RO \end{bmatrix} \end{cases} (R)\begin{cases} \begin{bmatrix} LC \\ IO \end{bmatrix} \\ \begin{bmatrix} LIC \\ O \end{bmatrix} \\ \begin{bmatrix} IC \\ LO \end{bmatrix} \end{cases}$$

$$(RL)\begin{bmatrix} IC \\ O \end{bmatrix}$$

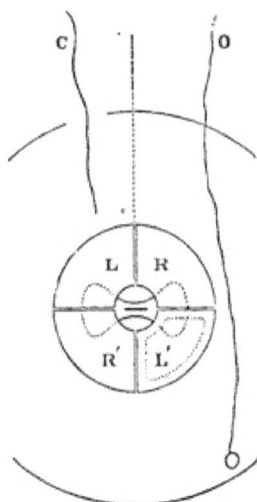

Fig. 19.

The facility afforded by the number of these arrangements for varying the sensibility of the instrument even to a moderate or slight degree without altering the adjustment of the fibres, will be found useful in some kinds of observations. For instance, if it be desired to observe the fluctuations of a varying potential, a degree of sensibility which throws the deflected image nearly to the extremity of the scale will cause the fluctuations to be twice as sensible and accurately read as if the deflection were only half as much, as they will bear the same proportion to the whole deflection in the two cases.

stead of being straight, is curved into the arc of a circle with the
mirror as centre. The actual angle of deflection can then be at
once calculated from the number of divisions of the scale passed
over by the light spot.

By substituting a lime-light for the paraffin lamp, and using
a larger scale, the instrument can be used for lecture purposes.

I have not given directions for adjusting and using the electrometer, as very full ones are sent out in pamphlet form with each
instrument.

CHAPTER VIII.

ON THE THEORY OF ABSOLUTE MEASUREMENT.

Of late years men of science have recognized the importance of being able to refer all measurements to a common system of units, or in other words, to render the numerical values of physical quantities independent of the particular instruments used to measure them.

We can easily imagine what confusion would be caused in the commercial world if there were no recognized standards of length and weight, but if every tradesman sold by an arbitrary weight and an arbitrary measuring-stick of his own.

An almost equally great confusion reigned in the scientific world till the system of absolute measure was developed by the labours of Gauss, Weber, Thomson, Fleeming Jenkin, Clerk Maxwell, Balfour Stuart, and others.

An electric force, for instance, was defined by one observer as one that required a torsion of 1000° of the thread of his torsion balance to neutralize it. The same electric force, however, if measured in another person's balance, would be equal to quite a different number of degrees of torsion.

Before Fahrenheit invented the first thermometer scale with absolute reference points, the same number on different thermometers represented quite different temperatures.

For the intensity of light no absolute scale yet exists.

But chiefly the defect was felt in the methods of measuring electric currents and resistances, and electro-magnetic effects.* Now, thanks partly to the requirements of the submarine telegraph companies, who have found it worth while to spend large sums of money on having these methods improved for them, the latter have been brought to a very high state of perfection indeed.

* See Part III.

ABSOLUTE UNITS.

The outline of the system of absolute measure is this :—

Certain units of mass, length, and time are chosen, and from them the derived **units** of density, area, volume, velocity, force, momentum, energy, &c., are deduced.

From these again are obtained units of electrification, potential, electric current, electric resistance, &c.

All these units, fundamental **and** derived, **are called** absolute **units.**

ABSOLUTE INSTRUMENTS.

Instruments **for the** measurement of electrical **and other** physical quantities whose scales are arranged to give the values of the quantities measured, **in** these absolute units are called absolute instruments. But **as** they **are** usually very expensive, very heavy, not portable, **and not** convenient to work with, smaller instruments with arbitrary **scales** are constructed.

Measurements of several values of **the** same physical quantity made by the small instruments **are** compared with measurements of the same values made by an absolute instrument, and a table of comparison between the scales is constructed.

This process **is** called *determining the constants* of **the smaller** instrument. **An** arbitrary instrument whose **constants have been** determined becomes an absolute instrument.

CHOICE OF UNITS.

The English Government has, for electrical and magnetic measurements, adopted the foot, the grain, and the second as the standards of length, mass, and time respectively, and this system is still used in the state observatories.

This system, though convenient for some purposes, is unsuited for scientific measurements, because the larger units of **each kind** are not decimal multiples of the smaller.

The ounce, for instance, is not a decimal multiple of the grain, or the pound of the ounce. The yard is not a decimal multiple of the foot, or the mile of the yard. Every time that measurements in the smaller or larger units have to be reduced into the larger or smaller, **a process of** division or multiplication has to **be** gone through.

C.G.S. Units.

To avoid this labour and source of error, scientific **men have** adopted **a decimal,** or so-called metrical system.

In this the second is still taken as the unit of time,

The centimetre is the **unit of** length, **and**

The gramme the unit **of** mass.

*The system of measurement based on these **units is called the*** *centimetre, gramme-mass, second system, or briefly the C.G.S. system.*[*]

Length.

The *centimetre* is **the one-hundredth part** of the metre.

The metre is **theoretically the** ten-millionth of **the quadrant** of the earth—that is, it is the ten-millionth of the result of a particular measurement of that quantity. Practically it is the distance at 0°C. between two lines engraved **near opposite ends** of a standard platinum bar preserved in the Paris observatory.

Should future **measurements of the earth's quadrant give a** number not exactly ten million times this distance, the standard metre will not be altered, but the earth's quadrant will be said to consist of more or less than ten million metres. The metre is equal to

39·370432 inches,

or to

1·09362311 yards,†

the standard metre being taken **as correct** at 0°C., the standard yard at

$$16\frac{2}{3}^{\circ} \text{ C. } (62^{\circ} \text{ F}).$$

The metre is thus divided :—

10 millimetres =	1 centimetre.	
100 millims. =	10 centims. = 1 decimetre.	
1000 millims. =	100 centims. = 10 decims. = 1 metre.	

---

[*] See **Units** and *Physical* **Constants,** by Prof. J. D. Everett, F.R.S. Macmillan and Co., 1879.

† For roughly **realizing what number in a** familiar measure **a** number expressed in the metrical system is equal to, we may put metre = 40 inches, decim. = 4 inches, centim. = $\frac{1}{5}$ inch, millim. = $\frac{1}{25}$ **inch.** The kilometre = 1000 metres nearly equals 5 furlongs.

MASS.

The *gramme* is theoretically the mass of one cubic centimetre of distilled water at 4°C.

Practically it is $\frac{1}{1000}$ of the mass of a piece of platinum preserved at Paris.

The masses of two bodies can be compared by simple weighing at the same place without knowing the force of gravity at that place, because the force of gravity is the same at the two ends of the balance, and therefore the weights are proportional to the masses.

The masses of two bodies at different places cannot be compared by sending the same spring balance to the two places, because, the force of gravity being different, the masses are no longer proportional to the weights.

Definition.—The mass of a body is the quantity of matter in it.

The weight of the body at any place is its mass multiplied by the force of gravitation at that place.

C.G.S. DERIVED UNITS.

DENSITY.

The density of a substance is the mass or quantity of matter per unit of volume.

In the C.G.S. system it is the number of grammes in a cubic centimetre. We now see the reason why the centimetre instead of the metre has been chosen as unit. It is because it makes the density of water unity; the adoption of the metre would have made it 1,000,000.

The densities of various substances then remain equal to their specific gravities—that is, to the ratio of their densities to that of water—and require no alteration to reduce them to absolute measure. The adoption of the metre as unit would have necessitated multiplying the specific gravities by 1,000,000 to give the densities.

We may here note that, of the two units of density and mass, we must assume one, and then can derive the other, but it is unimportant which.

The advantage in assuming a unit of mass is that masses can be compared more accurately than densities.

We can either say, as we have done, that the density of a

substance is the mass (expressed in arbitrary mass units, viz., the mass of a piece of platinum in Paris) in unit of volume, or we can say that the unit of mass is the quantity in a unit of volume of a substance (distilled water at 4°C.) whose density we arbitrarily take as unit.

The platinum unit of mass has been constructed, so far as possible, to make these two definitions identical; but if, owing to better determinations of the density of water, they should cease to agree, the unit of mass will not be changed, but a correction will be applied to the density of water. Meanwhile, the densities of other substances being determined experimentally by the ratio of those densities to that of water, the numbers expressing these ratios (the specific gravities) will remain unchanged, but will have to be multiplied by the new value of the density of water to give the true densities.

VELOCITY.

The velocity of any body moving with uniform velocity is the number of centimetres that it travels in a second.

The velocity at any instant of a body moving with variable velocity is the number of centimetres that it would have travelled in a second if it had gone on moving uniformly for a second with the velocity that it had at that instant.

A body moving at the rate of one centimetre per second has a velocity unity,* or *the unit of velocity is a velocity of one centimetre per second.*

ACCELERATION.

Let a point be moving with variable velocity; if the velocity is increasing, the motion of the point is said to be *accelerated.*

If the velocity is diminishing, the motion of the point is said to be retarded.

For convenience of calculation, both increase and diminution of velocity are called acceleration, but diminution is called negative acceleration.

It may easily be seen that this convention is legitimate, for if a moving point is being equally accelerated and retarded, its velocity will remain constant—the algebraic sum of the acceleration and retardation is zero—which shows that the retardation

---

* This, the unit velocity, is equal to about 40 yards per hour.

may be considered as a quantity equal in magnitude but opposite in sign to the positive acceleration. We may now define the unit of acceleration.

Uniform acceleration is the number of units of velocity by which the velocity of the moving point changes *in a second.*

Variable acceleration at any instant is the number of units of velocity by which the velocity of the moving point would change in a second, if for a second the rate of change had been that which it had at that instant.

The unit of acceleration is an *increase of velocity* of one centi-metre per *second.*

FORCE.

If a body, having mass, is moving with variable velocity, some force must be acting on it.[*]

The value of a force is the number of units of acceleration which it can produce on a unit of mass; [†] that is, it is equal to the number of units of velocity by which it can increase the velocity of a unit of mass in a unit of time.

The unit of force is that force which, acting on a mass of one gramme for one second, can increase its velocity by a velocity of one centimetre per second.

This unit of force is called a *dyne*. Now the velocity of a body falling *in vacuo* at Greenwich increases at the rate of 981·17 centimetres per second for each second of fall—that is, at Greenwich the earth acts on a body at or above its surface with a force of 981·17 dynes per unit of mass, or to hold up a gramme at Greenwich requires a force of 981·17 dynes.[‡]

To sum up—

The dyne or unit of force is the force which, if it acted for one second on a mass of one gramme, would, if the mass was previously at rest, give it a velocity of one centimetre per second, or, if it was previously in motion, in the direction of the force, would in that time alter its velocity by that amount. And a dyne is $\frac{1}{981\cdot17}$ of the force of gravity on a gramme at Greenwich.

[*] *Newton*, Laws I. and II.

[†] It also equals the number of units of *momentum* which it can produce in a unit of time.

[‡] Between the equator, **where it** is least, and the pole, where it is greatest, the earth's attraction varies by about $\frac{1}{186}$ of its whole mean value.

· Work.

The unit of work is called the erg ; it is the amount of work done by one dyne working through one centimetre—that is to say, it is equal to the work required to move a body through one centimetre against a force of one dyne. To lift one gramme one centimetre at Greenwich requires an expenditure of 981·17 ergs of work, and if at the top of the centimetre we let the body go, it is able in falling one centim. to do that amount of work.

C.G.S. applied to Electro-statics.

We now come to the C.G.S. system of measuring electro-static effects.

Quantity.

The electro-static unit of quantity is that quantity of electricity which at a unit distance will repel another equal quantity of the same kind of electricity with a unit of force—that is, in the C.G.S. system, it is the quantity of electricity which at a distance of one centim. will repel another equal quantity with a force of one dyne.

Now, if the unit be determined experimentally (we will describe the methods used later on), we see how, for instance, any torsion balance might be graduated. Let the fixed ball be charged with two units, and let the movable ball be made to touch it ; the balls will be then charged each with one unit of electricity and will repel each other. Let the torsion circle be then turned so as to bring them towards each other till they are at a distance of one centim.* The force between them is then one dyne, and it is exactly balanced by the torsion of the thread. The force exercised by the torsion of the thread at that particular position of the circle is then one dyne.

On repeating the experiment with charges of 3, 4, 5, &c. units, we shall get the readings of the circle corresponding to forces of 1½, 2, 2½, &c. dynes. On constructing a table of these results, unknown quantities of electricity can be at once measured in C.G.S. units by the instrument.† This and similar processes are called determining the constants of the instrument.

---

* The diameters of the balls are supposed very small compared with one centim.

† This is only intended as an illustration. The experiment could not be carried out exactly as here described.

POTENTIAL.

In the C.G.S. system, the unit potential *is **the** potential due to a unit of electricity **at** a distance of one centim.*

We have **defined** the *difference* of potential **between two** points, due to any given electrification, as the work **required to** be done to move a unit of electricity from one of these points to the other.

The unit difference of potential *is the difference of potential which there must be between **two** points for **one** erg of work to be required to move a unit of positive electricity from one to the other, the influencing electricity being supposed to retain its distribution unchanged.*

PLATE IV.—THOMSON'S LARGE ABSOLUTE ELECTROMETER.

CHAPTER IX.

ABSOLUTE ELECTROMETERS.

WE now come to the experimental methods of determining potentials and quantities of electricity in absolute units. The instruments used are called absolute electrometers, and are all the invention of Sir William Thomson.* Several forms of the absolute electrometer have been constructed. We shall only describe the latest.

THE NEW ABSOLUTE ELECTROMETER.—Plate IV.

The absolute electrometer consists essentially of two horizontal parallel plates insulated from each other.

In the centre of the upper one is a round hole which is almost filled up by a light disc hung by three delicate springs like carriage-springs. To the disc is attached a modification of the hair and dot arrangement described on page 41. The disc is so adjusted that, when the hair is sighted between the dots, it hangs so as exactly to fill the aperture in the plate—that is, the surface of the plate is continuous, except for the small annular space round the edge of the disc. The disc and the perforated plate, which is called the "guard-plate," are in metallic communication through the supports of the balance.

The suspended disc can be moved up and down by turning a micrometer screw, which moves the block to which the springs are attached.

The lower plate can also be moved up and down by a micrometer screw.

The simplest method of using the instrument is as follows:—

The upper screw is so adjusted that when no part of the apparatus is electrified, the disc hangs a little above the guard-plate.

* Report on Electrometers—Papers on Electrostatics and Magnetism Page 287—Sir Wm. Thomson (Macmillan), 1872.

The two plates (viz. the lower plate and the guard-plate and disc) being electrified by connecting them respectively with the two bodies whose difference of potential it is required to measure, the bottom one is moved up or down till the hair appears between the dots. This shows that, at the distance, which there then is between the plates, the attraction exactly balances the force, due to the springs, tending to move the disc upwards. The distance between the plates, the size of the suspended disc, and the force tending to move the disc upward, being all known, the corresponding difference of potential can be calculated by means of a mathematical formula.

The force tending to move the disc upwards is determined by finding what weights laid upon the disc will bring the hair to the sighted position when no part of the apparatus is electrified.

The use of the guard-ring is this: the formula is only applicable to the centre portion of an attracted plate—it is not applicable near the edges. The practical effect of the guard-plate is to give us a movable disc which is all centre and which has no edges. For the edges being those of the fixed guard-plate, the attraction on them does not affect the motion of the disc.

After a time Sir William Thomson found it better to replace this method of working by another not quite so simple. In the new method, each plate is insulated, and the upper one is charged to a high, and constant, potential. The lower plate is then connected alternately to the earth and to the body whose potential is to be observed. The difference of attraction in the two cases gives the difference of potential between the body and the earth—that is, the potential of the body.

To ensure the constancy of the potential of the upper plate, the latter is connected to a separate balance electrometer or gauge similar to that used as an accessory to the quadrant electrometer (fig. 15, p. 41). It is not necessary to know the potential to which it is charged, but only to know that it is constant. The accessory electrometer, or gauge, is called an *idiostatic electrometer.*

To guard against the effects of leakage, the glass case of the instrument is covered with tinfoil, and forms a Leyden Jar (page 61), in which a large quantity of electricity is stored.

Certain openings left in the tinfoil allow the interior works to be seen.

The potential of the charged plate is adjusted by means of a replenisher, like that used in the quadrant electrometer (fig. 16, page 42).

The suspended disc is protected from the effects of accidental induction by a metal cover, made in two halves, which are shown removed to the sides of the instrument.

A spiral wire, passing to a metal rod supported on a glass stem and projecting from the bottom of the instrument, enables connection to be made with the lower plate, and allows the latter to be moved up and down.

Two or three half-pint tumblers (not shown) stand inside the case, containing pieces of pumice-stone wetted with sulphuric acid, which absorbs moisture and keeps the air dry.

The whole instrument stands on three legs sufficiently high to enable the hand to be placed underneath to turn the screw of the lower plate.

The instrument is made by White of Glasgow.

To determine Potentials by the Absolute Electrometer.

It is shown in Sir W. Thomson's paper before quoted, and, though we can give his result here, we cannot give the mathematical proof, that

$$V = D \sqrt{\frac{8\pi F}{A}}$$

where V is the potential of the body under examination; D the distance between the plates; π the ratio of a circumference of a circle to its diameter—that is, $3\cdot1416$; F the electric attraction which is equal to the upward pull exercised by the counterpoise; A an area which is the mean of the areas of the suspended disc, and the hole in the guard-plate in which it hangs.

If, instead of being neutralized by the electric attraction, the upward force of the counterpoise, equal to F, were balanced by a weight which we will call W—that is, W grammes where W is a number different in different experiments—we have

Force = W multiplied by force of gravity at the place,

or in symbols

$$F = W.g.$$

and the formula becomes

$$V = D \sqrt{\frac{8\pi W.g}{A}}$$

and we remember that g at **Greenwich** is equal to **981·17** dynes.

In practice it was, however, **found** very difficult to determine D, the distance between the upper and lower plates, accurately. For this and other reasons the method of working was modified, as we stated on page 56.

Let D be the distance **when the lower** plate is connected to earth, and D' the distance when it is connected to the body under examination. Let V be the *difference* of potential between the upper and lower plate, when the latter is connected to earth; and V' the same quantity when it is connected to the body under examination.

Then. as the potential of the upper plate is constant, the difference of the potentials of the lower plate, when connected **to** earth and **to** the charged body, will be

$$V - V' *$$

The formula then becomes

$$V - V' = (D - D') \sqrt{\frac{8\pi W \cdot g}{A}}$$

and $V - V'$ is the difference **of** potential between the earth and the charged body, and $D' - D$ **is the** *difference* between the two readings of the screw **of the lower** plate, and can be determined with **perfect accuracy without** a knowledge of **the actual distances.**

The Portable Electrometer.—Plate V.

Sir William Thomson has also invented a *portable electrometer.* Its scale **is** graduated by direct electrical comparisons with the absolute instrument.

Details of the instrument are shown in Plate V.

The arrangement is somewhat similar to that of the New Absolute Electrometer. The "attracted disc" is square, and with its guard-plate forms an arrangement $(h, f, \text{fig. 2})$ exactly **like the guage** of the Absolute Electrometer or of the quadrant.

---

* For let potential of upper plate be A, and that of the lower, B and B' in the two experiments respectively, then—

$$V = A - B, \qquad V' = A - B'$$

and $V - V' = B' - B$, and the difference between B and B' is the potential we want to measure.

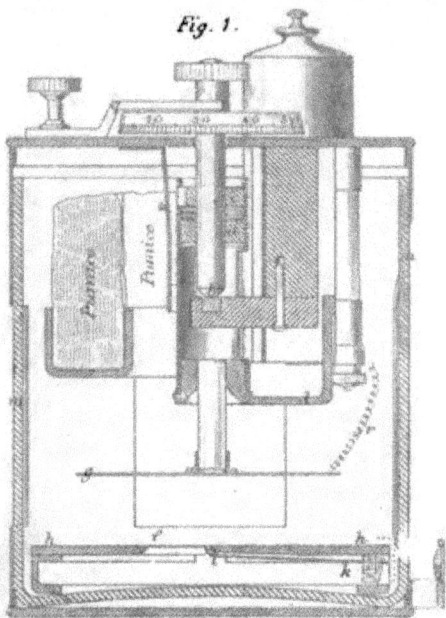

Fig. 1.

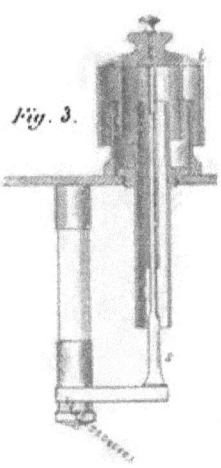

Fig. 3.

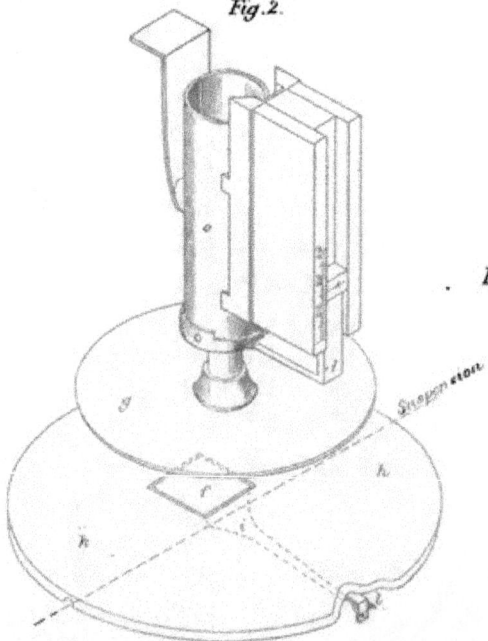

Fig. 2.

Portable Electrometer
Figs. 1 & 3
⅔ Full size

It is inverted and placed at the bottom of the instrument, and is attracted *upwards* by the movable plate (*g*, fig. 1).

The case forms a Leyden jar, the inside of which is connected to the gauge and guard-plate.

To use the instrument the jar and gauge are highly charged.

The movable plate is connected to earth, and screwed up and down till the gauge balances at the sighted position.

The reading of the moving plate having been taken, it is connected to the body whose potential is to be measured, and moved till the disc again balances.

The difference between the two readings of the movable plate gives the difference of potential between the earth and the body, in units of the instrument, assuming the potential of the gauge and guard-plate to have remained constant throughout the experiment.

To test this a second earth reading should be taken.

If the two earth readings differ widely, the experiment must be rejected. If they exactly agree, the potential has remained constant, while, if they differ slightly, their mean will not be far from what would have been the earth reading at the time when the charged body was tested.*

The value of the arbitrary units of the instrument are determined, once for all, by a series of measurements of the same potentials by it and by the Absolute Electrometer.

The mechanical arrangement by which the plate is moved up and down is very ingenious. The plate *g* is fixed to a circular vertical brass stem, which is pressed into two brass V's by a spring (fig. 2). This ensures its always moving parallel to itself.

It is moved up and down by a screw with a convex steel head (*b*, fig. 1), which rests upon an agate plate. On the screw are two nuts (*a* and *c*, fig. 1) ; one is fixed to the tube carrying the plate *g*, and the other is only prevented from revolving inside it. They are forced apart by a spiral spring. This prevents what is called in mechanics " back-lash" or " lost time"—that is, it ensures that the longitudinal motion shall be reversed at the same time as the circular.

The screw is kept against the agate by pressing the milled-

---

* This assumes that the loss is uniform, and that the time from the first earth reading to the charge reading is equal to that from the charge reading to the second earth reading.

head with the finger. The number of whole turns of the screw are read on the scale (fig. 2), while fractions of a turn are read on the circular head attached to it (fig. 1).

The details of the gauge are shown in figs. 1 and 2. *h h* is the guard-plate, *f* the balanced plate, *i k* the pointer, and *l* the hair and dot arrangement. The tinfoil is removed to allow the hair to be seen, and a screen of fine wire gauze is substituted to guard against electrical influence. '

The Umbrella, fig. 3, is a guard to protect the rod communicating with plate *g* from accidental induction. It can be slid up **and down** as required; when lowered, it makes communication between *g* and the outer case.

Some *pumice* just moistened with sulphuric acid serves to **keep** the air dry. It must be carefully dried in an oven once a month.

The instrument is chiefly used for observations of atmospheric electricity. It is not fitted for extremely accurate work.*

REASON FOR CHARGING THE DISC.

In both the portable and in the absolute electrometer the screw plate would have to be moved equally for a given difference of potential, **however** feebly the disc was **charged**, or even if it was not charged at all.

The effect of highly charging the disc is that, although the difference of attractions is the same, the total attractions are greater, and slight irregularities in the actions of the springs would not cause so great an error.

* With great care potentials equal to half a Daniel's cell may be measured. With ordinary practice and care, potentials equal to from two to three cells. The quadrant electrometer measures to about $\frac{1}{30}$ of a cell, but can be adjusted to measure $\frac{1}{100}$, and it takes 1100 cells to produce a spark $\frac{1}{30}$ inch in length.

CHAPTER X.

THE LEYDEN JAR.

WE have shown that when two oppositely charged conductors, separated by an insulator, are brought near together, they will attract each other.

If we can arrange two conductors of large surface, and place them very near together, separated by a rigid insulator, we shall have all the conditions for a very powerful attraction.

These conditions are satisfied in the " Leyden Jar," which in its most ordinary form consists of a wide-mouthed bottle of hard white glass (fig. 20), coated inside and out with tinfoil. The tinfoil stops a few inches from the mouth of the bottle. The bottle is closed by a lid of hard wood, in the centre of which is a brass rod with a knob at the top. A chain hangs from the lower end of the brass rod and touches the inside tinfoil.

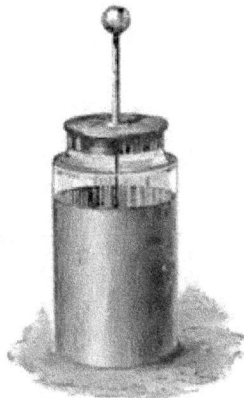

Fig. 20.

The inside tinfoil can be charged with positive electricity by placing the knob of the jar near the conductor of a glass electric machine, and working the latter till sparks pass to the knob. When sparks refuse to pass any longer, it shows that the inner tinfoil is charged to almost the same potential as the conductor of the machine. This (+) charge acts inductively *through the glass*, and induces a (−) charge on the inside of the outer tinfoil, and a (+) one on its outside.

If the outer conductor be connected to earth, the side furthest from the inner conductor may be considered to be removed to an indefinite distance, and only (−) electricity remains in the outer tinfoil.

As the outer tinfoil entirely surrounds the inner one, the induced (−) charge on it will be equal to the inducing (+) one.

We have then two opposite charges of electricity spread over large surfaces, and separated only by the thickness of the glass.

These two charges attract each other very strongly, and, the moment a path is opened by which they can come together, they will combine with great violence.

Experiment.—Take a charged jar and a pair of "discharging tongs." These latter consist of a jointed conductor attached to an insulating handle or handles. Touch the outside coating of the jar with one end of the tongs, and bring the other near the knob. There are now two strains going on; one is the strain of the glass which is constant, and the other the strain of the air between the knob and the tongs, which is increasing as they are brought nearer together. Meanwhile, as the thickness of the stratum of air between the knobs diminishes, its mechanical strength—that is, its power of resisting disruptive discharge—decreases. At last a point is reached when the air is no longer strong enough to resist the stress or straining force, and the electricities burst through it and combine with a flash and a report. Immediately after this has occurred, the jar is found to be completely discharged.

Residual Charge.

If, after being discharged, the jar be left to itself for a few minutes, it will be found to have again acquired a small charge. This second charge is called the "Residual Charge." With a half-gallon jar it is usually sufficiently powerful to produce a spark visible in daylight.

The phenomena of residual charge can only be explained by supposing the induction passing through the glass to be a state of strain of the particles of the glass. On this hypothesis we suppose the glass in the charged jar to be very much strained, but not to be perfectly elastic. On the tinfoils being discharged —that is, on the removal of the straining force—the particles of glass instantly fly back almost, *but not quite,* to their normal unstrained position. For a moment we then have the tinfoils discharged, but the glass in a slightly strained state. In the course of a few minutes the glass slowly recovers from this residual strain.

Thus, while the inner tinfoil has remained insulated, a change has occurred in the electrical arrangement of the particles of glass near it. *The state of strain has altered.*

Now, in the ordinary phenomena of induction, the effect of altering the state of strain of an insulator (by bringing a charged body near it) is to induce a charge on any adjoining conductor.

In the present case the residual charge is produced by the change from a more to a less strained state, which takes place in the glass by virtue of its elasticity.

MECHANICAL ILLUSTRATION.

The following mechanical illustration will help us to obtain a clearer view of what takes place.

Experiment.—Take a strip of **gutta-percha** about 6 inches **high,** 2 wide, and ½inch thick, and fix its lower end to a table or in a vice so that it stands up vertically. Now bend the upper end down with the finger.

The **gutta-percha** represents the glass of the jar, and the pressure of the finger the straining force of the **charge.**

Now suddenly let go the gutta-percha—that is, discharge the jar. It will fly to a position nearly, *but not quite,* vertical, and will rest there for a moment.

If the finger be placed on it, there will be no pressure felt for a second or two.

This represents the instant when the jar is completely discharged. In a few seconds more the gutta-percha will **begin** to again press on the finger, and, **on being** released, **will fly** to the vertical position—that is, the recovery from the **residual strain** will have developed a residual **pressure or charge.**

EFFECT OF TAPPING.

A further proof that the phenomena of the Leyden jar are the effects of strain is found in the fact that *any mechanical agitation of the particles of the glass, which enables them to move more freely over one another, hastens the development of the residual charge.*

The following experiment is described by Dr. **John Hopkinson** in the *Phil. Trans.* 1876, page 489, and was successfully **shown** by the present writer to a large audience at the Royal Institution on January 23, 1879.

Experiment.—A **glass bottle,** intended for making a small Leyden jar, about 3 inches high, is used, but is not coated with tinfoil.

The conductors **consist of** strong sulphuric acid. The jar is half filled with it for the inner conductor, and stands in a glass dish of it which forms the outer one. The glass dish is insulated by being laid on a slab of india-rubber.

The connections with the **jar** are made **by platinum wires** dipping into **the acid.**

To charge the jar, the outer acid is connected to earth by means of a platinum wire attached to the nearest water-pipe, and the acid in the bottle is connected to a friction electric machine—about a 15-inch **plate is a** suitable size. The jar is charged, **by** working the **machine for some two or** three minutes, **and is** then insulated.

It is then discharged **by** means of a piece **of platinum** wire, **of the shape** ∩, attached to an insulating handle.

The inside and outside **acids are** kept **connected** for about ten to twenty seconds.

The connection having **been broken,** the inside and outside are connected respectively by **platinum** wires to **the** quadrants of an Elliott-pattern electrometer.*

The **return of the** residual charge **should cause** a slow and steady motion **of the light-spot. If** the light-spot moves too fast, **the two wires must be pressed together between** the finger **and thumb** for ten or fifteen seconds more, **so as to** further discharge **the** jar, and this process must be repeated until a convenient speed is obtained.

As **soon** as this has **been done, the** edge of the jar may be **tapped** smartly with a piece **of hard** wood (the handle **of a small** hammer, **for** instance). *The pace at which the light-spot moves will be at* **once** *trebled.*

This **shows** that the residual charge **is** appearing about three times as fast as when the jar was undisturbed.

This experiment is perfectly explained on the hypothesis **of** induction being a state of strain of **the** insulator; for we know that if **any partly** elastic body be **distorted,** and is slowly re-

* See page 37.

covering its normal shape by virtue of its elasticity, any tapping or jarring will, by enabling the particles to slide more freely over each other, greatly hasten that recovery.

When it is desired to show this experiment to an audience, a lime-light, and a scale about eight feet long, placed at a distance of about twelve feet from the electrometer, should be substituted for the ordinary scale and lamp. With this arrangement and a suitable lens, a clear disc of light, about two inches in diameter, is seen on the scale, and, when the jar is undisturbed, it should move at the rate of about three inches per second.

If it is desired to repeat the experiment, the electrometer can be discharged, and the light-spot brought back to the zero by holding the platinum wires together for an instant.

In order to check the oscillations of the needle of the electrometer, the wire which leads from it down to the acid should have a piece of thin sheet platinum, about $\frac{1}{2}$ inch wide by $\frac{1}{4}$ inch high, attached to its lower end below the surface of the acid.*

SUPERPOSITION OF DIFFERENT CHARGES.

In the same paper Dr. Hopkinson has shown that if the jar be charged alternately + and — for various periods, the residual charge will be first — and then will become +, showing that, as he expresses it, "the charges come out in the reverse order to that in which they went in."

The experiment succeeds even when the charge has been reversed three or four times.

The importance of this result lies in the fact that, while it would be impossible to conceive of alternate "actions at a distance," remaining superposed on each other in the glass, it is yet quite possible to conceive that alternate strains of its particles should be so superposed.

---

* The following is a convenient way to attach it. The platinum is cut

into the shape and the lower end of the wire is bent into a

hook of the shape The end *a* of the platinum is then bent over and
hung on the hook.

F

MECHANICAL ANALOGY.

On October 4, 1878, **Dr. Hopkinson** communicated a paper to the Royal Society,* "On the Torsional **strain** which remains in a glass fibre after **release from** Twisting **Stress.**" In the paper he tests experimentally **an** assumption which **had** been previously made **by** Boltzmann, that the principle **of** superposition, which we have **just** mentioned, **is** applicable **to the** mechanical strains induced **in the glass** fibre **by** torsion. **In his** experiments **Dr.** Hopkinson **found** that the assumption **was a correct** one, **and that, if various** twisting **forces** were applied, the **effect** of each, separately, was stored up **in** the fibre, and could **be** detected **in** turn **after the** release of the latter from the twisting stress.

Thus a complete analogy **is** established between the twisting **of a** glass fibre and the phenomena of residual **charge;** and we can have **no** doubt that **the** electrical effects are due to mechanical strains of the **insulator.**

EFFECT OF TEMPERATURE.

In the *Phil. Trans.* **1877, vol.** clxvii., Dr. Hopkinson **has shown** that all the phenomena of the Leyden jar are affected by changes of temperature.

This is another proof that the effects are due to **a** strain of the glass, for the rigidity and other mechanical properties of the glass are **all** affected by temperature; but we could not imagine the temperature of the glass affecting a "direct action at a distance" passing **through** it.

ANALOGY BETWEEN LEYDEN JAR AND STRAINED BEAM.

Mr. Ayrton and Mr. Perry have published an experimental investigation, in which they have compared the mechanical straining **of** beams with the absorption and return of electricity in extremely well-insulated Leyden jars, each experiment lasting many days. They have found that the curves expressing **the** rate of return from the strained state are precisely **similar in the** mechanical and electrical cases.†

OTHER CONDENSERS.

A Leyden jar is sometimes called a " *Condenser,*" because it

* Proc. Roy. Soc., vol. xxviii. 1878, page **148.**

† See Chapter **XXXI.**, and Proc. Roy. Soc., vol. xxx. 1879-80, page 411.

was formerly supposed that the strong electrical effects observed were due to the condensation of an electric fluid or fluids.

Condensers are made in many other forms besides the jar form which we have been describing.

They sometimes consist of a flat plate of an insulator coated on both sides with tinfoil. The tinfoils are made smaller than the insulator in order that they may be efficiently separated.

The two tinfoils act as the inner and outer coatings of a Leyden jar.

Sometimes a condenser, of very large surface, is formed by placing a great number of alternate plates of insulator and tinfoil together.

In this case the 1st, 3rd, 5th, ... tinfoils are connected together, and correspond to one coating of the Leyden jar, and 2nd, 4th, 6th, ... are connected and correspond to the other.

The insulator in these large condensers is sometimes mica and sometimes paper which has been dipped in melted paraffin wax.

Messrs. Clark and Muirhead's great condenser, which has been constructed for "duplexing" the direct United States' cable, contains 100,000 square feet, or more than 2 acres, of tinfoil, and fills 70 boxes each 2 feet by 1 foot 6 inches and 7 inches thick.

CAPACITY.

The capacity of a condenser is measured by the quantity of electricity of unit potential which it can contain, or, in other words, it is equal to the charge divided by the potential.

It is found that up to a certain limit it gets less as the thickness of the insulator gets greater.

When a metal sphere is hung up in the centre of a room, it forms the inner coating of a condenser, of which the air is the insulator, and the walls, floor, and ceiling, the outer coating connected to earth.

When the room is very large, the insulator can be considered as being so thick that variations in its thickness do not affect the capacity.

The capacity then only depends on the size of the sphere, being larger when the radius is increased.

It can be shown mathematically that the capacity of such a sphere is simply and directly proportional to its radius, and the

68   *Electro-Statics.*

units have been so chosen that *the capacity of an isolated* spherical conductor *is numerically equal to its radius*

C.G.S. Units.

The C.G.S. unit of **capacity is** (in Electro-static measure) the capacity **of** an isolated spherical conductor of **one centimetre** radius.

CAPACITY OF TWO CONCENTRIC SPHERES.

It can be proved mathematically that **the** capacity of a condenser consisting of two concentric spheres* **is** equal **to the** product of their radii divided by the thickness of the air space between them; **that is, if r and** r' **are the radii** of the inner **and** outer spheres respectively, **then**

$$\text{Capacity} = \frac{r\,r'}{r' - r}.$$

---

* **The inner one being charged,** and the outer connected to earth.

CHAPTER XI.

SPECIFIC INDUCTIVE CAPACITY.

Introductory.

IF Electric Induction were a "direct action at a distance," we should expect that it would be transmitted equally through all insulators. One of the strongest arguments for supposing it to be a strain of the particles of the insulator is found in the fact that different insulators transmit it with very different strengths.

We have defined the capacity of a condenser as the quantity of electricity of unit potential which it will contain.

This quantity depends :—

(1) On the size of the conducting plates; for a large plate will hold more electricity than a small one.

(2) On the strength of the inductive action between each square centimetre of the opposed plates.

The last quantity depends on two things :—

(1) On the thinness of the insulator; for the nearer the two conductors are to each other, the stronger will be the inductive action between them ; and,

(2) *On the specific power of the substance of which the insulator is composed, of receiving and transmitting that electric strain which we call induction.*

This power is called the Specific Inductive Capacity of the substance.

The specific inductive capacity of dry air, at the ordinary pressure and temperature of the atmosphere, is taken as the standard and called **unity**, and the capacities of other substances are compared with it.

Definition.—Let there be two condensers, one (*a*) having air as its insulator, and the other (*b*) having any other substance; and let the condensers be precisely similar in all other respects, then *the ratio of the capacity of* (*b*) *to that of* (*a*) *is called the Specific Inductive capacity of the substance which forms the insulator of* (*b*).

We see that the Specific Inductive Capacity of any substance is a quantity which expresses the ratio between the power of transmitting electric induction **possessed** by that substance and by air.

In actual determinations of Specific Inductive Capacity, it is not always necessary to make the two condensers precisely similar, as when the form and dimensions of the air condenser are known, we **can** calculate, from experiments on it, what would have been the capacity of an air condenser similar to the condenser under examination.

Definition.—Insulators across which electric action takes place, such as the insulators of condensers, are called " *dielectrics*," from the Greek διὰ, across.

The determination of the specific inductive capacities of different dielectrics is one of the most important branches of electrical research.

It is of the very greatest commercial importance, for in submarine telegraphy the number of words per minute which can be transmitted through the cable—that is, the gross receipts of the company—depend, in a great measure, on the lowness of the specific inductive capacity of **the insulator of the cable.**

In choosing between different insulators, it is therefore absolutely necessary to have an accurate **knowledge of** their capacities.

In the theoretical study of electricity it is of even more importance, for, as we shall show in Part IV. of this book, the theory which **at** present offers us the fairest hopes of some day finding out exactly what electricity is, requires certain relations between the specific inductive capacities and the refractive indices of transparent dielectrics. It is only by accurate determinations of specific inductive capacities that this theory can be experimentally tested.

We shall therefore give a very full account of all the determinations of specific inductive capacity which have been made up to the present time.*

* July, 1882.

We must premise that the experiments are of an extremely difficult and complicated nature, and that the numbers obtained for the same substances by different observers differ very widely.

The reasons of the differences are not yet fully understood, though many causes of error have lately been discovered and eliminated; but we cannot yet say for certain whether the true specific inductive capacities of the same substances are or are not altered, by changes in the strength of the electrification, by the duration of the charging, and even by molecular changes occurring spontaneously, from month to month, in the dielectrics themselves.

Nevertheless, it is only by a careful study of what has been already done that we can hope to evolve some order out of the uncertainty in which existing experiments have left us.

EXPERIMENTS OF CAVENDISH.*

Between 1771–81 Cavendish measured the capacities of various condensers by comparison with certain other standard condensers which he called "*trial-plates.*"

The capacities of these trial-plates were themselves determined in absolute measure, by comparison with a globe 12·5 inches in diameter suspended in the middle of his laboratory.

An actual air condenser was constructed, consisting of two brass plates, 8 inches diameter, placed parallel to each other, with only air between them, at distances which, in the different experiments, varied from ·910 to ·259 inch.

From his experiments, Cavendish calculated the ratios of the capacity of glass and other substances to that of air; that is what we now call their specific inductive capacities.

In comparing the capacities of two condensers, we must remember that, if we connect them both simultaneously to the *same* source of electrification, the charges which they will receive will be directly proportional to their capacities, and, therefore, to know the ratio of their capacities, it is only necessary to measure the ratio of their charges.

The first portion of Cavendish's experiments consisted of a comparison between the capacities of his condensers and those of

* See *The Electrical Researches of the Hon. Henry Cavendish, F.R.S.*, written between 1771–81, edited by J. Clerk Maxwell, F.R.S. Cambridge, Univ. Press, 1879. Pages 144—188.

the various "trial-plates" which he constructed as **arbitrary** standards.

His condensers consisted of flat plates of glass, shellac, and beeswax, on opposite sides of which circular discs of tinfoil were pasted.

The trial-plates were glass with tinfoil discs. They were made in a set of ten, of sizes to give different "computed charges"— that is, to be of different capacities.

The capacity of the smallest being called unity, there were in the set—

3 plates of capacity	1
3 „ „ „	3
3 „ „ „	9
1 „ „ „	27

By suitable combinations of these ten plates, **a condenser**, of any capacity from 1 to 66, could be constructed.

One of the trial-plates was made with a sliding conductor for slightly varying the capacity **as a** fine adjustment.

It consisted **of a** brass plate which could either be laid entirely on the tinfoil, **or** slid so **as** to project over **its** edge, and thus practically coat a larger portion of the glass.

The method of comparison was what is **called** a "zero method" —that is, instead of the difference between the condensers and the trial-plates being measured, **the latter** were adjusted till they exactly equalled the condenser—that **is, until** there was *no difference.*

The **following method was** used for adjusting the trial-plate to equality with **the** condenser under examination :—

In fig. 21, let B be the condenser, T the trial-plate. One coating of each was connected to earth, and **the** other coatings together to the knob of a charged Leyden jar.

By being connected to the **same** jar, **both** condensers were charged to the **same** potential, and therefore the charges x and y in each were simply proportional to their capacities.

After the charging had gone **on for** about two seconds, the connections were altered to the arrangement of fig. 22 by a convenient mechanical contrivance, consisting of a system of strings and wires supported **on** a wooden frame.

This we see sends the charges $+y$ and $-x$ to **earth, but com-**

Lines the charges $+x$ and $-y$. A pith ball electrometer (page 31) is attached to the connecting wire.

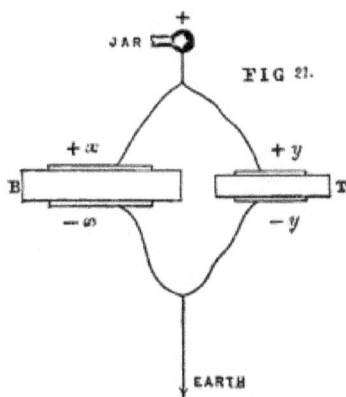

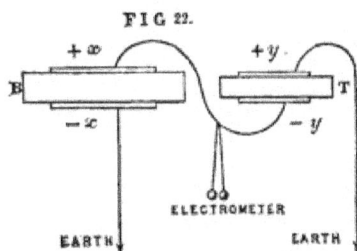

FIG 21.

FIG 22.

If the capacities of B and T are equal, we shall have the charges equal—that is, $x = y$ and $(+ x) + (- y) = 0$—that is, the algebraic sum of the equal and opposite charges will be zero, and the pith balls will not diverge.

If the pith balls diverge with $(+)$ electricity, it will show that the capacity of B is the largest, and that of T must be increased.

If they diverge with $(-)$ electricity, it will show that the capacity of T is the largest and it must be decreased.

In practice it was found better to have two trial-plates, T and T', and by separate experiments to make T of just less capacity than B, and T' of just more. The mean

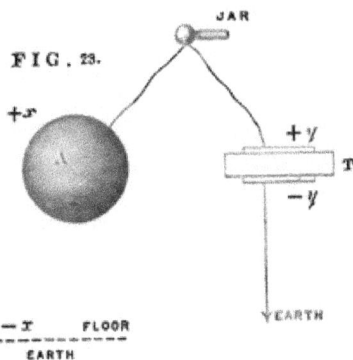

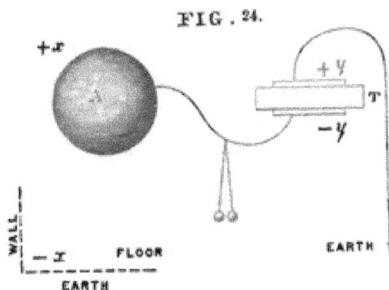

FIG. 23.

FIG. 24.

of the capacities of T and T' was taken as the capacity of B. This was because the electrometer was more sensitive when slightly diverged than when collapsed.

COMPARISON OF TRIAL-PLATE AND SPHERE.

The comparison of the trial-plate and sphere was made in an exactly similar way.

The sphere A (figs. 23 and 24) took the place of the upper coating of the condenser B, and the walls and floor of the room, which can always be considered to be connected to the earth, replaced the lower coating. The arrangement for charging was then that of fig. 23, and the charges $+x$ and $-y$ were combined by connecting, as in fig. 24.

The student should compare these figures with figs. 21 and 22.

SPREADING OF ELECTRICITY.

Some difficulty was experienced in determining the diameter of the charged surface, for it was found that the electricity *spread* a little all round the edges of the tinfoil, and that therefore in the calculations it was necessary to suppose the tinfoil to be somewhat larger than it really was.

The amount of correction to be applied was determined by comparing whole discs of tinfoil with others in which there were numerous slits cut. The "whole" and "slit" plates were arranged to be of the same actual area, but of course the slit one had a very much greater length of edge. The ratio of the capacities of the "slit" and "whole" condenser gave the ratio of the areas of the charged surfaces—that is, the ratio of the areas of the slit tinfoil and electrified space round its edges, to the whole tinfoil and electrified space round its edges.

From these data the actual area of electrified surface could be determined.*

* Let A be the area of each tinfoil.
     a the electrified area round the whole one.
     · *b* „   „   „   „   „ slit one.
     M the ratio of the charges.
     N the ratio of *a* to *b*—that is, the ratio of the lengths of the edges.
Then A, M, N are known by experiment, and we have to find *a*.
     We have

$$M = \frac{A + a}{A + b} \text{ or } M (A + b) = (A + a),$$

Cavendish found that the effect of the instantaneous spreading of the electricity was about the same as if the tinfoil plates had extended $\frac{7}{100}$ inch in every direction, when the thickness of the glass was about $\frac{1}{8}$ inch, and about $\frac{9}{100}$ when the thickness was about $\frac{1}{15}$.

THE EXPERIMENTS.

By the use of a mathematical formula,* Cavendish computed what ratio the charge of each condenser should have borne to that of the globe, if air had been substituted instead of glass, as the dielectric. He then compared the charges experimentally.

The ratio of the "observed charge" to the "computed charge" is, if the formula is correct, the ratio of the capacities of glass and air—that is, the specific inductive capacity of glass.

In order to test the formula, the sphere was compared with an "air condenser" consisting of two brass plates fixed parallel to each other with only air between them. It was then found that the charge of the plate condenser was about $\frac{1}{10}$ greater than it should have been, according to the formula. This difference was no doubt caused by the action of the walls, &c., of the room. The theory requires an infinitely large room; the actual laboratory was only about 16 feet square.

$$\text{or } (M - 1)\, A = a - M\, b,$$

but $b = \frac{a}{N}$; substituting we have

$$(M - 1)\, A = a \left(1 - \frac{M}{N} \right)$$

$$\text{or } a = \frac{(M - 1)\, A}{1 - \frac{M}{N}}$$

* According to theory, the capacity of an air condenser consisting of flat parallel circular plates of equal diameters is equal to that of an isolated globe, whose diameter equals the square of the radius of either plate divided by twice the distance between them.

In modern language, the capacity of an isolated sphere is equal to its radius, and the capacity of a plate condenser where r is the mean of the radii of the plates, and a their distance apart, is equal to that of a sphere of radius x, where

$$x = \frac{r^2}{4\,a}$$

In the following tables the "computed charge" has, **if I**
rightly understand the paper, been calculated on the supposition
that no such difference exists. The excess of the observed
charge over the computed charge is due, I think, partly to the
difference of specific inductive capacity between air and glass,
and partly to the difference in the shape of the condensers. The
ratio of observed charge to computed charge is the proportion
in which the capacity would have been increased by the substitution of glass for air, and by the substitution of a flat condenser
for a spherical one.

If my view of the paper is correct, then the observed charge,
divided by the computed charge, is the specific inductive capacity
of glass *when that of air is taken as* 1·1.

To obtain the true specific inductive capacity of glass—that is,
the capacity of glass when that of air is taken as 1·0—we must
diminish the numbers in the proportion of 11 to 10 and say,

$$\text{Spec. Ind. Cap.} = \frac{10}{11} \cdot \frac{\text{observed charge}}{\text{computed charge}}$$

The following is the result of Cavendish's experiments on
plates of glass, shellae, and bees-wax, as given in Prof. Maxwell's
edition, and I have added a column in which the above correction
is applied :—

TABLE OF GLASS PLATES.

Dielectric.	Thick-ness.	Dia-meter.	Ditto cor-rected.	Specific gravity.	Com-puted charge.	Ob-served charge.	Ob-served charge by com-puted charge	$\frac{10}{11}$ of last column = Specific Inductive Capacity.
Flint glass ground flat	·2115	2·23	2·37	3·279	3·32	26·3	7·93	7·21
Ditto a thinner piece	·104	2·215	2·385	3·284	6·84	52·3	7·65	6·95
Plate glass D	·127	2·85	3·02	2·752	8·68	71·9	8·01	7·28
" " W	·172	3·435	3·565	2·787	9·34	74·8	8·01	7·28
" " G	·1848	3·575	3·725	2·973	9·38	75·5	8·05	7·32
" " N	·106	2·12	2·29	2·682	6·18	51·4	8·31	7·55
" " O	·106	2·505	2·675	2·514	8·44	75	8·90	8·08
" " Q	·076	2·065	2·245	2·504	8·29	76·5	9·23	8·39
Crown Glass	·0682	3·495	3·675	2·537	24·76	211·3	8·54	7·76
Ditto another piece	·0659	3·43	3·61	2·532	24·72	208·7	8·44	7·67
Crown glass ground	·07	2·035	2·215	$\Big\{ 2·685$	{ 9·76	76·5	8·73	7·93
Part of same piece	·0683	3·54	3·72		{ 24·96	215·1	8·62	7·83
Mean of the 10 pieces used in former experiments				2·678	...	...	8·22	7·47

PLATES OF OTHER SUBSTANCES.

Dielectric.		Thickness.	Diameter.	Computed charge.	Observed charge.	Observed charge by computed.	$\frac{10}{11}$ of last column = Specific Inductive Capacity.
Gum Lac . . .		·125	4·23	17·89	80	4·47	4·09
Mixture of rosin and bees-wax. Plate	1	·1645	3·75	3·63	13·5	3·72	3·38
	2	·192	3·355	7·22	25·2	3·49	3·17
	3	·103	4·247	21·80	69	3·15	2·86
	4	·103	4·525	21·85	78·9	3·18	2·89
	5	·103	1·79	3·89	13	3·34	3·03
Dephlegmated bees-wax. Plate	1	·303	3·79	5·90	24·5	4·16	3·78
	2	·120 ·	3·525	12·95	46·1	3·56	3·23
	3	·063	2·74	14·90	50·5	3·39	3·08
Plain bees-wax .		·119	3·475	12·69	51·3	4·04	3·67

The coatings of all these plates were circular.

In computing the charge of the glass plates, the diameter of the coating was corrected on account of the spreading of the electricity as in the fourth column, the electricity being supposed to spread ·07 of an inch if the thickness is ·21, and ·09 if the thickness is ·08, and so on in proportion in other thicknesses. But no correction was made in computing the charges of the other plates, as the experimenter was uncertain how much to allow.

Cavendish gives the following account of the method used for making the plates of lac, rosin, and bees-wax. " I first cast a round plate of the substance, three or four times as thick as I intended it should be, and rather thinner near the edges than in the middle, taking care to cast it as free from air bubbles as I could.

" I then heated it between two thick flat plates of brass, till it was become soft, and then pressed it out to the proper thickness by squeezing the plates together with screws. In order to prevent its sticking to the brass plates, I put a piece of thin tinfoil between it and each plate, and I found the tinfoil did not stick to it so fast but what I could get it off without any danger of damaging them.

" The heat necessary to melt shellac is so great as to make it froth and boil, which makes it impossible to cast a plate of it free from air bubbles. The plate mentioned in the preceding table was as free from them as I could make it. It contained, however, a great quantity of minute bubbles, but no large ones.

" Bees-wax melts with a heat of about 145° Fahr. **If it is then**
heated to a degree rather greater than **that** of boiling water, **it**
froths very much, and seems to lose **a good deal** of watery matter ;
and if it is kept at this heat till it has ceased frothing, it will
then bear being heated to a much higher degree without frothing
or boiling.   Bees-wax thus prepared **I call** dephlegmated.

" In order that the plates of dephlegmated bees-wax should all
be equally so, I dephlegmated some bees-wax with a pretty considerable heat, and suffered it to cool and harden, and out of this
lump **I** made all three plates, taking care in casting them not to
heat them more than necessary.

" I used the same precautions **also** in casting the plates of a
mixture of rosin and bees-wax ; the **proportion of** the rosin to the
bees-wax was forgot to be set down.

" What are called in the table the 4th and 5th plate of rosin
and bees-wax are in reality the same plate as the 3rd, only with a
smaller coating."

Cavendish goes on to say, " It appears from these experiments,
first, that there is a very sensible difference in the charge* of
plates of the same dimensions according to the different sort of
glass they consist of, the charge of the plates O and Q, which
consisted of greenish foreign plate glass being the greatest in
proportion to their computed charge of any, next to them the
crown glass, and **the** flint glass being the least of all.

" Secondly, the charge of the lac plate is much less in proportion to its computed charge than that of any glass plate, and
that of **a** plate of bees-wax, **or of** the mixture of rosin and bees-
wax **still** less.

" But it must be observed that there is a very considerable
difference between the three different plates of dephlegmated
bees-wax in that respect. The same thing, too, obtains in the
mixture of rosin and bees-wax.

" **As** the proportion of the real charge to the computed [the
specific inductive capacity] is greater in the thick **plates than**
the **thin ones, one** might be inclined to think that this was owing
to the electricity being not spread uniformly. But as the
difference seems to be greater than could well proceed from that
cause, I am inclined to think that it must have been partly owing
to some difference in the nature of the plates. Perhaps it may

* Here the charge is proportional to the Specific Inductive Capacity.

have been owing to some of the plates having been less heated, and consequently having suffered a greater degree of compression in pressing out than the others.

"The piece of ground crown glass mentioned in the first of the foregoing tables was made out of a piece of crown glass about ¼ of an inch thick, and ground down to the thickness mentioned in the table, care being taken by the workman to take away as much from one side as the other, so that the plate consisted only of the middle part of the glass.

"My reason for making it was that as there appears to be a considerable difference in the charge of different sorts of glass, it was suspected that there might possibly be a difference between the inside of the piece and the outside, and if there had, it would have affected the justness of the experiments with the ten pieces of glass ground out of the same piece.

"But by comparing the charges of the plates of crown glass with those of the two other pieces of crown glass in the table, there does not seem to be any difference which can be depended on with certainty.

"The experiment indeed would have been more satisfactory if the piece of ground glass and the pieces with which it was compared had been all made out of the same pot. But as it would have been difficult procuring such pieces, and as I have found very little difference in the specific gravity of different pieces of crown glass, and as I am informed it is all made at the same glass house, I did not take that precaution."

CYLINDRICAL CONDENSERS.

Cavendish also made some experiments "On the charges of such Leyden vials as do not consist of flat plates of glass."

"These experiments were made with hollow cylindrical pieces of glass, open at both ends, and coated both within and without with pieces of tinfoil surrounding the cylinder in the form of a ring, the breadth of the ring being everywhere the same, and the inside and outside coatings being of the same breadth, and placed exactly opposite to each other. Only as the inside diameter of the two thermometer tubes was too small to admit of being coated in this manner, they were filled with mercury by way of inside coating.

"The thickness of the glass was found by suspending the

cylinder by one end from a pair of scales, with its axis in a vertical position, and the lower part immersed in a vessel of water, and finding the alteration of the weight of the cylinder according as a greater or less portion of it was under water.

"The result of the experiments is contained in the following table :—

	Mean Thickness.	Mean outside semi-diameter.	Length of coating.	Specific gravity.	Outside diameter by thickness.	Computed charge.	Observed charge.	Observed charge by computed.
Part of a jar of flint glass .	·084	1·02	4·4	3·254	19·3	85·9	717	8·35
A cylinder of do.	·0704	·645	9·50	3·281	9·2	87·1	680	7·46
Thermometer—								
Tube I. . .	·094	·14	11	3·096	1·5	11·0	80·2	7·31
„ II. . .	·130	·16	15·5	3·243	1·24	11·1	80·7	7·26
Cylinders of {1	·045	·50	7·16	2·865	11·3	77·2	754	9·77
green bottle {2	·069	·53	8·55	2·664	8·8	78·6	690	9·00
glass {3	·078	·48	7	2·865	6·2	40·8	355	8·65

He adds that :—

"The lengths of the coating here set down are the real lengths. But in computing the charges of the white jar and cylinder and the three green cylinders, these lengths were increased on account of the spreading of the electricity according to the same supposition as was used in computing the charges of the flat plates.

"But in computing the charges of the thermometer tubes no correction was made, as I was uncertain how much to allow, but as the length of their coatings is so great, this can hardly make any sensible error."

In these experiments we must take the observed charge by computed charge to be the specific inductive capacity, as Cavendish made no accurate experiments on the capacity of cylindrical air condensers.

EFFECT OF TEMPERATURE.

Cavendish also experimented on the effect of temperature on the specific inductive capacity of glass, and found that at 295° Fahr. the glass conducted pretty freely, but at 305° Fahr. much faster. He also concluded that the capacity was considerably increased when the glass became conductive, but that until then there was no sensible difference.

EFFECT OF VARIATION OF CHARGE.

Cavendish also made some experiments to determine whether

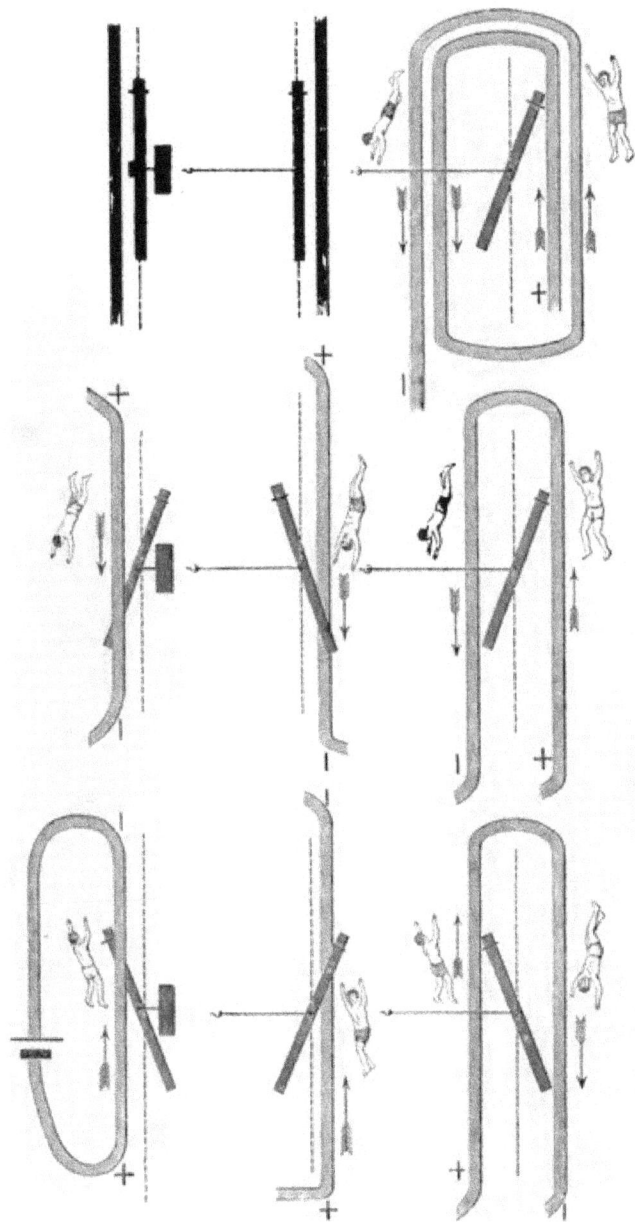

PLATE XXI.—ACTIONS OF CURRENTS ON MAGNETS.

specific inductive capacity varied at all with electro-motive force, but within the limits of his experiments he could detect no difference.*

EXPERIMENTS OF FARADAY.

Before describing the more modern experiments, we must mention that the researches of Cavendish remained unpublished until September, 1879, and that neither Faraday nor any other of the experimenters, whose work we are about to notice, even knew of their existence at the time when their experiments were made.

On November 30, 1837, Faraday communicated to the Royal Society a paper† "On Induction," in which he announces the discovery (or, as we now must say, the re-discovery) of Specific Inductive Capacity, and describes methods of measuring it, and the results obtained by him for glass and other substances.

His apparatus consisted of two exactly similar spherical condensers of the shape shown in fig. 25.

The inner coating of each consisted of a brass ball (A) 2·33 inches in diameter, and the outer one of a hollow brass sphere (B) whose internal diameter was 3·57 inches. This left a space of 0·62 inch between the outer and inner surfaces.

The inner ball was supported by an insulating stem (*l*) of shellac, inside which a wire passed up from the ball to a little knob (*a*) outside.

The outer sphere was divided at its equator, and the two halves could be separated for the introduction of various dielectrics.

Different gases could be admitted by means of the tap R.

Fig. 25.

The space between the balls could be filled with different insulators, and we see that by keeping one apparatus full of air, and filling the other with any other substance, we should have

---

* It is possible that it varies with some substances, and not with others, just as the distortion due to mechanical stresses would be proportional to the stress with some substances, but not with others.

† *Exp. Res.* 1161, vol. i. p. 360.

two condensers exactly similar, except in the nature of their dielectrics.

In practice it was found more convenient to make the dielectric in the form of a *hemi*-spherical cup, and having measured its effect to calculate what the effect of filling the whole space would have been.

The following was the method of working :—The two condensers were set side by side with their outsides connected to earth, and one of them was charged with electricity. The potential of the charge was measured by means of Coulomb's torsion balance.*

The insides of the two condensers were now connected together and the potential measured again.

We know that, if two condensers be connected together, the charge will divide between them in direct proportion to their capacities; and also that, if half the charge be taken away from any condenser, its potential will be one-half what it was before.

If therefore the second condenser is of the same capacity as the first, it will take exactly half the charge, and *the potential, after division, will be one-half what it was before.*

If however the second condenser were of greater capacity than the first, it would take more than half the charge, and the potential, after division, would be less than half what it was before.

If the second condenser were of less capacity, the potential would be more than half.

THE EXPERIMENTS.

Some preliminary experiments were made, with both condensers full of air only, in order to test their equality, and then a hemispherical cup of shellac was placed in one apparatus, and the other left full of air.

In order to compensate the effects of leakage, the air apparatus and the shellac apparatus were alternately charged first in different experiments. As the effect of leakage would in the one case be to make the capacity of the lac apparatus too high, and in the other too low, the mean of the two experiments would be very nearly unaffected by it.

The results of the first two experiments were—

* Page 33.

Ratio of capacity of lac condenser } { Air charged first 1·55
to that of air condenser. } { Lac charged first 1·37

After many other experiments, Faraday came to the conclusion that the ratio of the capacity of the shellac condenser to that of the air condenser was almost exactly 1·5.

From this result Faraday calculated that, if the shellac had entirely filled the condenser, the ratio would have been at least 2·0—that is, that the specific inductive capacity of shellac is 2·0.

Faraday experimented on **other** substances, and the following is a table of his results :—

Dielectric.	Specific Inductive Capacity.
Shellac	2·0
Sulphur	2·24
Glass	More than 1·76.
Oil of Turpentine . .	More than unity.

GASES.

A very long and careful series of experiments was made on gases. No less than twenty-five pairs of gases were compared. Comparisons were also made between air, damp and dry, hot and cold, and at high and low pressures, but in no case was Faraday able to detect any difference at all.

OTHER EXPERIMENTS.

After Faraday's **experiments**, the subject of Specific Inductive Capacity **remained untouched** till 1871, when **Messrs.** Gibson and Barclay **published a paper** on it. Since that **date** hardly a year has passed without some addition to our knowledge.

MESSRS. GIBSON AND BARCLAY'S EXPERIMENTS ON PARAFFIN.[*]—
Plates VI. and VII.

The **chief** instruments employed were—
(1.) The Quadrant Electrometer, **White** pattern, (page 39).
(2.) The Platymeter.
(3.) The Sliding Condenser.

THE PLATYMETER.

The Platymeter consists essentially of two condensers of equal

* *Phil. Trans.* 1871, p. 573.

capacity of cylindrical form, the inner conductors of which are in metallic connection.

It is shown in Plate VI. fig. 1, where it will be seen that the two inner conductors form one cylinder, cc; the outer conductors are the two short cylinders, pp, $p'p'$.

METHOD OF WORKING.—Plate VII., fig. 3.

Let one pair of quadrants of the electrometer be connected to earth, and the other pair to the conductor, cc, common to both condensers of the platymeter.

Suppose one of the outer cylinders, pp, to be charged positively. It will induce a charge opposite to its own on the cylinder cc, and one similar to its own on the quadrants. Let the quadrants, still connected to the cylinder, be connected to earth, and then again insulated. Then the positive will have escaped, and the negative, being attracted by the positive on the outer cylinder, will produce no effect on the electrometer as long as the induction from the outer to the inner cylinder remains unaltered. So there will be no deflection.

Now let the two equal and similar outer cylinders, pp, $p'p'$, be connected.

The inductive effect of pp on cc is reduced to one-half, for pp has lost half its charge; but, at the same time, $p'p'$ having gained half the charge of pp, will exercise an action exactly equal and similar to that of pp. So the sum of the actions of the two half-charged cylinders, pp, $p'p'$, on cc, the inner cylinder, will be exactly the same as previously was the action of the one fully charged cylinder pp, and so there will be still no deflection of the electrometer.

Thus we see, as long as the capacity of the two condensers, viz., pp and part of cc, and $p'p'$ and part of cc are equal, there will be no deflection.

If now we repeat the experiment with two other condensers added to pp and $p'p'$ respectively, there will still be no deflection if the two new condensers are equal; but if they differ, the needle will move.*

---

* If the platymeter had not been used, the two new condensers could only have been compared by charging them to equal and opposite potentials by means of a very large battery. With the platymeter an electric machine or Leyden jar can be used.

Fig 1.

Longitudinal Section of Platymeter.

Fig 1.

Section through A.B.

Fig 2.

Section through C.D.

Fig 2.

Section through A.B.

Scale of Inches

Fig 2. Longitudinal Section of Sliding Condenser.

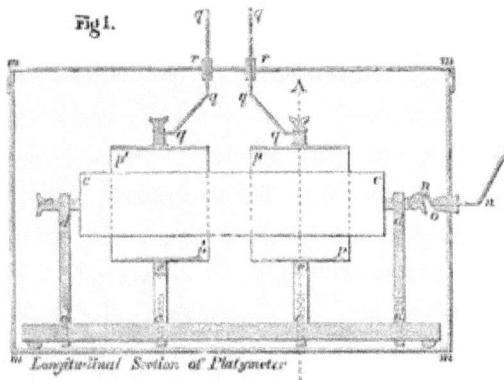

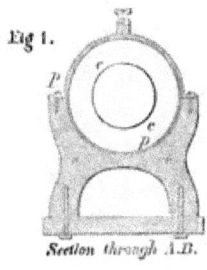

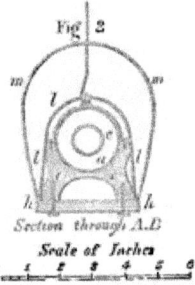

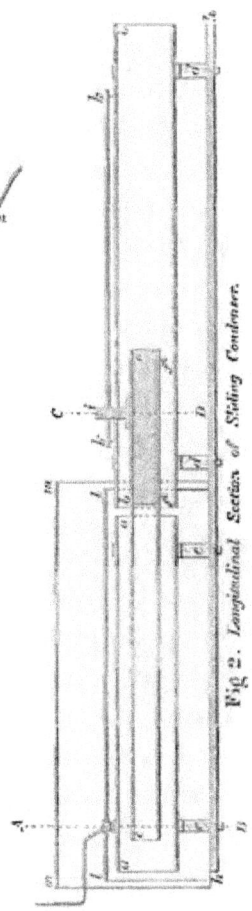

PLATE VI.—GIBSON AND BARCLAY ON SPECIFIC INDUCTIVE CAPACITY.

Fig 3.

Scale of Inches

Sectional Plan showing Connections

Earth

Fig 4.

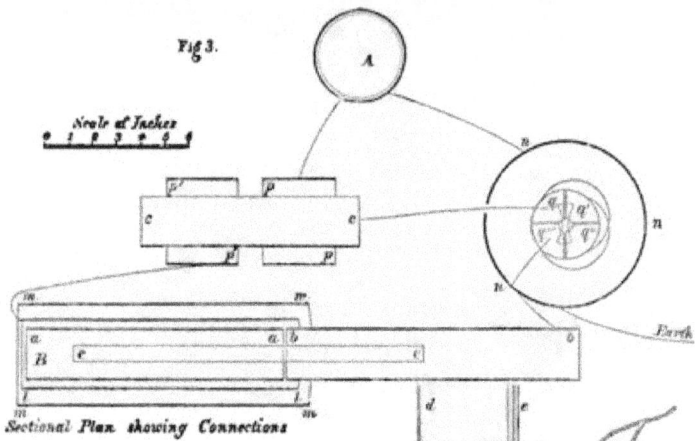

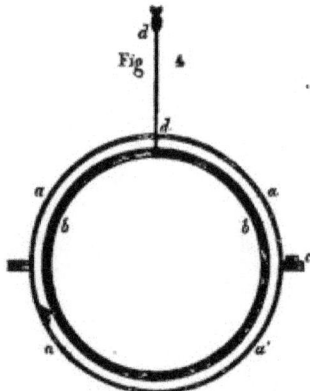

Section of Spherical Condenser

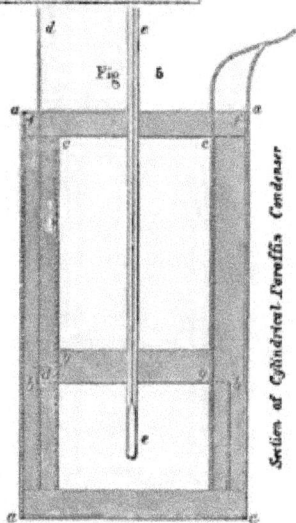

Fig 5.

Section of Cylindrical Paraffin Condenser

Scale of Inches

PLATE VII.—GIBSON AND BARCLAY ON SPECIFIC INDUCTIVE CAPACITY.

If a condenser whose capacity is a fixed quantity be connected to pp, and one of variable capacity to $p'p'$, then, if we vary the latter till there is no deflection, we shall know that we have adjusted the two new condensers to equality.

A condenser of fixed capacity and whose capacity could be easily calculated, and having air for its dielectric, was made by fixing a metal ball of known radius inside another (Plate VII., fig. 4) whose radius was also known.*

This was attached to one of the outer cylinders pp, and a condenser of variable capacity, called "the sliding condenser," was attached to the other, $p'p'$ (Plate VII., fig. 3).

The sliding condenser having been adjusted to equality with the known fixed condenser, its scale was read and noted.

The fixed air condenser was then removed, and another one, whose dielectric was the dielectric under examination, was substituted for it, and the sliding condenser again adjusted to equality with it.

It is clear that the ratio of the two capacities of the sliding condenser is the same as the ratio of the capacity of the air condenser to that containing the dielectric which is being experimented on.

The Sliding Condenser.

The sliding condenser is shown in Plate VI., fig. 2.

It consists of a fixed insulated tube, $a\,a$, and another, $e\,e$, which can be slid inside it. When $e\,e$ and $a\,a$ form the two conductors of a condenser, the capacity of the condenser is greater or less according to whether more or less of $e\,e$ is inside $a\,a$.

The tube $b\,b$ merely acts as a support for the sliding tube $e\,e$, and carries the scale by which the position of $e\,e$ is read. It has no electrical use, and is kept connected to earth.

The shaded part of $e\,e$ is weighted, and it slides on four brass feet, ff, inside $b\,b$.

The value of one scale division of this was calculated—that is, the alteration of capacity which would be caused by moving it one division. This can be calculated when we know the radii of the two cylinders.† The whole capacity of the sliding condenser could not be accurately calculated.

* Page 68. ·

† Let r be the radius of the inner cylinder, r' that of the outer one, l the

It could then be determined how many scale divisions were equal to the capacity of the spherical condenser—that is, how much the sliding condenser would have had to be moved to bring the needle back to zero, if, after equilibrium had been established, the capacity of the spherical condenser had been suddenly doubled.

This being known, *the ratio of the capacity expressed by one scale reading to the capacity expressed by another is at once known.*

THE PARAFFIN CONDENSERS.

Condensers were prepared having paraffin instead of air as the dielectric.

One consisted of a flat circular brass box, inside which a brass disc was supported on three ebonite pins; melted paraffin, being poured in, solidified, filling the spaces between the disc and the box. Owing to some defects in this condenser, another was constructed, consisting of two concentric tubes (Plate VII., fig. 5), the space between which was filled with paraffin.

The paraffin condenser being substituted for the air condenser, the sliding condenser was adjusted to equality with it.

A comparison of the two scale readings of the sliding condenser gave the ratio of the capacities of the air and paraffin condensers, and from this result the specific inductive capacity of the paraffin could be calculated, allowing for the difference in shape and size of the air and paraffin condensers.

RESULTS.

The results obtained, after making various corrections, were,— From experiments on the disc and box paraffin condenser—

Specific inductive capacity of paraffin = 1·975 ;

from experiments on the cylindrical paraffin condenser—

Specific inductive capacity of paraffin = 1·977.

The latter value is adopted by the authors.

The density of the paraffin at 11°C. was ·9080.

---

length of one scale division, Δc the change of capacity caused by moving the cylinder one scale division, then it can be shown mathematically that

$$\Delta c = \frac{l}{\log \frac{r'}{r}}$$

Boltzmann's Experiments.[*]—First Method.

In 1874, Prof. Boltzmann published an account of some determinations of specific inductive capacity. One of his condensers consisted of two circular parallel plates insulated from each other, and mounted in vertical planes on a sliding stand which was graduated so that their distance apart might be known.

In this various dielectrics could be placed. The distance of the plates was usually so adjusted that they did not touch the dielectric.

When a plate was inserted, the capacity of the condenser was determined experimentally, and from it the specific inductive capacity was determined by the following formula:—

Let m be the distance apart of the plates, n the thickness of the dielectric, and K its specific inductive capacity.

Then, when there is only air between the plates, the capacity will be inversely proportional to m, the distance between them.

If we now fill the space between the plates with the dielectric of specific inductive capacity K, we shall produce the same increase of capacity as if we had reduced their distance from m to $\frac{m}{K}$, and the ratio of the two capacities would be m to $\frac{m}{K}$ —that is, K to 1.

When however the dielectric has only a thickness n, it displaces a plate of air of thickness n, and substitutes for it a plate whose effect is the same as if the stratum n had been diminished to a thickness $\frac{n}{K}$.

The capacity will then be increased as if the plates had been brought nearer by a distance n, and then separated by a (less) distance $\frac{n}{K}$, and it will be inversely proportional to

$$m - n + \frac{n}{K} \dagger$$

and the ratio of the capacities will be in the inverse ratio of m to the above expression, thus—

$$\frac{\text{Capacity of Condenser with air only}}{\text{Capacity of same condenser with dielectric}} = \frac{m - n + \frac{n}{K}}{m} \qquad . \quad (1)$$

---

* *Carls Repertorium*, x. p. 109, and *Wiener Sitzungsber.*, Bd. lxvii. part ii. p. 17.

† Compare page 111.

As m, n, and ratio of capacities, are known quantities, K can easily be calculated from this equation; for if R be the ratio of the capacities, it gives us, by ordinary algebraical transformation,

$$K = 1 + \frac{m}{n} (R - 1). \quad \ldots \quad \ldots \quad (2)$$

The fact that the capacity of the air condenser was inversely as the distance of the plates, and was not affected by any disturbing cause, was carefully tested experimentally.

THE ELECTROMETER AND BATTERY.

A Thomson Electrometer and 18 Daniell cells * were used.

THE EXPERIMENTS.

The comparison of the capacities of the condensers was made as follows :—

(1) The battery was connected directly to the electrometer, and the deflection noted.

(2) One plate of the condenser with only air in it was attached to the battery, and the other to earth.

(3) The charged condenser plate was separated from the battery and connected to the electrometer.

The deflection was now smaller than in (1), and from the difference the ratio of the capacity of the condenser to that of the electrometer could be calculated,† for the charge divides between them in the inverse ratio of their capacities.

All three processes were now repeated with a dielectric inserted, and this gave ratio of "dielectric" condenser to electrometer. We now know ratios of capacities of

$$\frac{\text{Air condenser}}{\text{Electrometer}} \text{ and } \frac{\text{Dielectric condenser}}{\text{Electrometer}}.$$

The ratio of these two ratios is the ratio of the capacity of

* Chapter XVIII.

† For the ratio of the deflections is inversely as capacity of electrometer to capacity of electrometer and condenser together.

Writing D E and D C for the electrometer and condenser deflections respectively, and C E and C C for their capacities, we have

$$\frac{D E}{D C} = \frac{C E + C C}{C E},$$

$$\text{or } \frac{C C}{C E} = \frac{D E}{D C} - 1.$$

That is, ratio of capacities equals inverse ratio of deflections minus unity.

the air condenser to that of the "dielectric" condenser—that is the quantity R in the formula (2), page 88.

The experiments were carried out as follows :—

Part 1.—A wire DC, fig. 26,* was connected to the electrometer, and three other wires, I. II. III., connected as shown, were arranged so that they could be made to touch D C when required.

Before commencing an experiment, I. was lowered, which discharged the electrometer. Then I. was raised and II. lowered. This charged the electrometer direct from the battery, and the deflection was noted as "battery result."

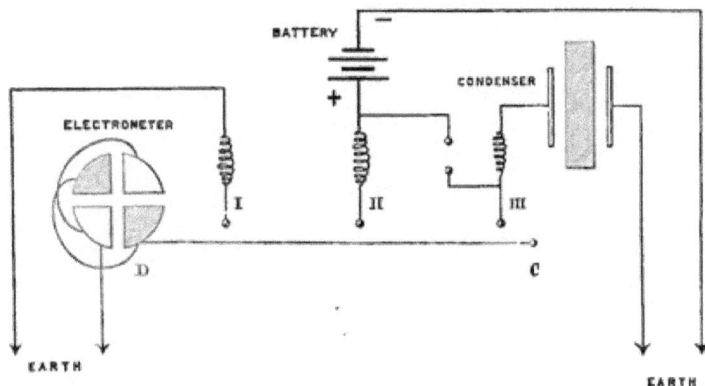

Fig. 26.

Then II. was raised, and I. lowered and raised, and the apparatus was ready for the next part of the experiment.

Part 2.—The dielectric being removed, III. was raised above its mean position so as to charge the condenser. III. was then lowered so as to connect condenser to electrometer.

The deflection was noted as "air-condenser result."

Finally, Part 2 was repeated with the dielectric inserted, and the deflection noted as "dielectric condenser result."

In order to make sure that there was no accidental electrification of any part of the apparatus, II. was connected to earth, and all the movements repeated. When this could be done without any deflection being produced, it showed that there was no accidental charge.

---

* The parts of the wires above the spirals are supposed to be fixed.

DURATION OF THE CHARGING.

The condenser and the electrometer remained in contact long enough to allow the deflection to be read, and then both were discharged by sinking I.

For quicker experiments an arrangement of springs (not shown) was attached to III., by which it was made to touch an earth wire instantly after both its top and bottom contact.

To test the effect of absorption, experiments were made, first as quickly as possible, so that the contacts only lasted a fraction of a second, and then slowly, so that each operation lasted from one to two minutes, but no difference in the results could be perceived when good insulators were used.

Again the condenser was charged for a long time, then discharged for a moment, and then connected to the electrometer, but no residual charge could be observed.

QUICKSILVER CONDENSER.

To try the case where there was no air in the condenser—that is where the plates were in metallic contact with the dielectric—a "quicksilver condenser" was constructed.

The plate was laid in a dish containing mercury, and mercury poured on its upper surface, a paper rim being attached round the edge to prevent the mercury running off. The results were found to agree with the former ones.

IMPERFECT INSULATORS.

Finally, some imperfect insulators, such as glass, gutta-percha, stearine, &c., were tried, but these were so much affected by changes in strength and duration of the charging that no consistent results could be obtained.

RESULTS.

The following determinations were made :—

Dielectric.							Specific Inductive Capacity.
Ebonite.	.	.	.	.	.	.	3·15
Paraffin	.	.	.	.	.	.	2·32
Sulphur	.	.	.	.	.	.	3·84
Resin	.	.	.	.	.	.	2·55

BOLTZMANN'S SECOND METHOD.[*]

On July 24, 1873, Professor Boltzmann read before the Vienna

[*] Experimental-Untersuchung über die electrostatische Fernwirkung dielectrischer Körper. Wiener Sitzungsber., Bd. lxviii. part ii. p. 81.

Academy a paper on "the electrostatic Far-working of dielectric bodies," in which he investigates the question of specific inductive capacity from an entirely new point of view.

If an unelectrified metal ball be suspended by a long fine thread, and be attracted by a fixed charged metal ball, the amount of attraction will depend only on the sizes of the balls, their distance apart, and the strength of the charge ; and can be calculated by known mathematical methods.

If a ball of some insulator, such as sulphur, be substituted for the suspended metal ball, the attraction will depend not only on the distance and sizes of the balls and on the charge, but also on the specific inductive capacity of the sulphur.

From a comparison of the attractions on a metal and on a sulphur ball, under the same circumstances, the specific inductive capacity of the sulphur can be calculated.

The present paper consists of accounts of a series of experiments in which the attractions exercised on metal balls were compared with those on balls of various dielectrics.

The first method used was not very accurate, but as it was simple we will describe it in order that we may more easily understand the more complicated methods which were developed out of it.

The "metal ball" and the "sulphur ball" were each 7 millims. diameter, and were suspended 90 millims. apart by threads of unspun silk, each 2 metres long (fig. 27).

A charged metal ball, which we shall in future call the "working ball," of 26 millims. diameter, was placed exactly half-way between them. Both the suspended balls were electrified by induction, and attracted with forces depending on the intensity of the "Far-working" on each ball.

In order to measure the attraction, a scale was placed behind the threads, and the positions of the latter on it observed by means of two microscopes.[*] As the deflections were very small, the attractions were taken to be proportional to the number of divisions passed over by the threads.

Fig. 27.

---

[*] Inserting the "working ball" set the threads swinging. The limits of swing were observed, and their mean taken as the deflected position.

The ratio of the deflections was written **down as**

$$\frac{\text{Force on metal ball}}{\text{Force on sulphur ball}},$$

and was called E.

As the attraction of the metal ball was always the greatest, E was always greater than unity.

In order that the weights and volumes of the two balls might be exactly equal, they were both made from the dielectric under examination, but the one chosen for the metal **ball** was either gilt, or covered with tinfoil.

The first experiment was **made** with sulphur, and the result was—

$$E = \frac{\text{Attraction of metal ball}}{\text{Attraction of sulphur ball}} = 1\cdot90.$$

Professor Boltzmann shows that K, the specific inductive capacity, can be calculated from E by the following mathematical formula : [*]—

$$E = \frac{K + 2}{K - 1}$$

which gives

$$K = \frac{E + 2}{E - 1}.$$

Thus the capacity of sulphur, as calculated from the first rough experiment, would be

$$E = 1\cdot90$$

$$K = \frac{3\cdot90}{\cdot90} = 4\cdot3 \ldots$$

The experiments **were** repeated with the working ball charged positively and negatively alternately.

These experiments were affected by several causes of error. The chief defect of the method is due to the fact that its accuracy depends entirely on the accurate placing of the working ball **midway between** the suspended **balls,** and this is an extremely **difficult** adjustment to make.

---

[*] Prof. Boltzmann first calculates the attraction of a ball of any dielectric, and obtains an **expression** involving K. He then obtains **the** attraction of a metal ball by putting $K = \infty$ in his formula. Dividing one expression by the other, he finds that **the** ratio E is equal to

$$\frac{K + 2}{K - 1} (1 + \text{a very small quantity}).$$

See Appendix to this chapter, p. 135.

The suspended balls were roughly made and were much disturbed by currents of air. The silk threads by which they were suspended were also not very good insulators.

For these reasons Professor Boltzmann modified his method of proceeding as follows :—

IMPROVED METHOD—INSTRUMENTS.

The balls, instead of being simply hung by threads as in fig. 27, were attached alternately to the same point at the end of the arm of a torsion balance, and the attractions calculated from the respective deflections.

The torsion apparatus is shown in fig. 28. Its arm consists of a stiff wire E F G H, to one end of which the sulphur or metal ball L is hung by two fine threads attached at the points G H.

At the other end of the arm is a mirror, *s*, in which the deflection is observed by means of a scale and telescope.[*]

The arm is suspended by a bifilar[†] arrangement from the points A B. M is the working ball which attracts the ball L, and it is connected to a Leyden jar.

To protect the arm itself from the action of the electrified bodies, it is enclosed in a triangular conducting case of gilt paper, connected to earth.

The case is shown dotted in fig. 28, and solid in fig. 29.

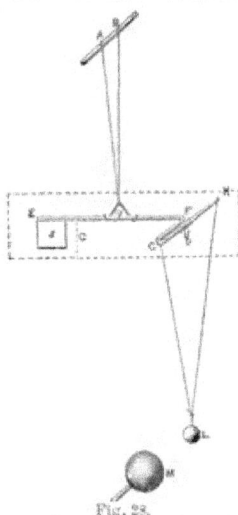

Fig. 28.

The apparatus is mounted on a stand, as in the right-hand side of fig. 29.

As the metal and sulphur balls are placed alternately and not simultaneously at L, it is necessary to know that the potential of the working ball remains constant throughout the experiments. For this purpose a separate electrometer, shown on the left of fig. 29, is used. A metal ball, P, at the end of a suspended metal rod, is attracted by a fixed ball, R, in metallic connection with the working ball M.

* This method of observation differs slightly from the "lamp and scale" method already described. See Chapter XIV.

† See pages 37 and 43.

P' is merely a counterpoise. The rod P P' is connected to earth by a wire ending in a platinum plate J, which dips into a

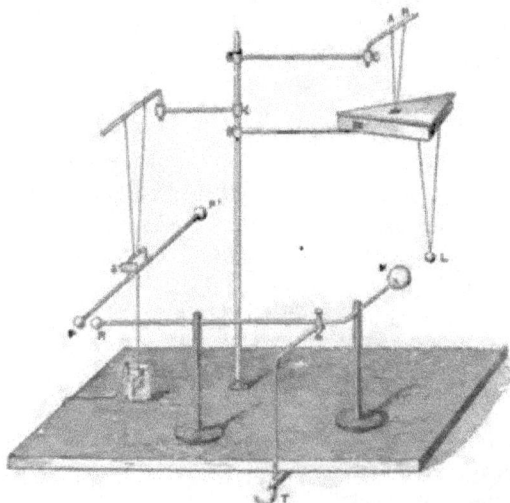

Fig. 29.

jar of dilute sulphuric acid. Another wire connects the jar to the nearest water-pipe. The plate J helps to check the oscillations of the arm P P'.

The deflection of P P' is read by means of a mirror S'.

M and R can be charged with electricity by means of the wire T.

To protect the suspended arms from currents of air, the whole apparatus is, when in work, covered with a glass case with a plate glass front, through which the mirrors can be observed.

It was not found necessary to keep the potential of M absolutely constant, as, from the deflection of P, small differences in the potentials in different experiments could be calculated and allowed for.

In the actual experiments the deflection of the arm E F G H was first observed with a sulphur ball at L, then with a metal one, and then with a sulphur one again alternately.

A door in the side of the glass case enabled the balls to be changed, and a little clamp at C (fig. 28) held the arm steady while the exchange was being made.

In case of the ball and the mirror not quite balancing, the rod could be slid through the hook at *g*.

REGULATION OF THE CHARGE.

In order to regulate the charge, the arrangement of fig. 30 was used.

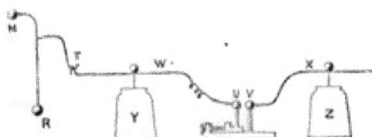

Fig. 30.

A Leyden jar, Y, was attached to the balls M R, and was connected by the wire W to a "discharging electrometer," U V.

DISCHARGING ELECTROMETER.

This consists of two insulated metal balls, U V (fig. 30), whose distance apart can be adjusted by a screw. If sparks be passed from one to the other, it is found, other circumstances being the same, that for any given distance, the quantity of electricity which passes in one spark is approximately constant.

METHOD CONTINUED.

A charged jar Z is connected to the ball V by a wire X, and one or more sparks allowed to pass. The potential to which the balls M R are charged is proportional to the number of sparks, for it is equal to the quantity of electricity which has passed, divided by the capacity of the jar Y and the wires and balls M, R, T W, U.

SHORT CHARGES.

For trying the effect of charges of very short duration, the arrangement of fig. 31 was used. The wire W was broken at *a b*, and the ends *a b* supported mercury cups, which could be connected by means of a bent wire *c*. *c* was attached to a pendulum, and made and broke the contact alternately at short intervals.

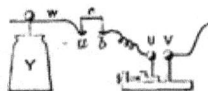

Fig. 31.

First contact was made at *a b*, and the balls charged by passing a spark between U V; then contact was broken at *a b*, and U connected to earth.

Then contact was made at *a b*, which discharged the whole apparatus. All these operations took rather less than $\frac{1}{10}$ second.

RAPID ALTERNATE CHARGES.

In order to alternately charge the balls **M and** R with (+) and (—) electricity, a wire *g* (fig. 32) was **led from** them to a

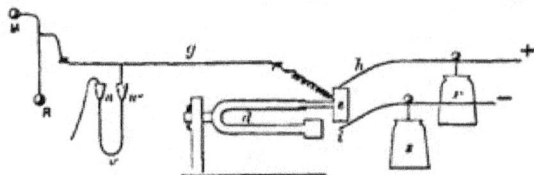

Fig. 32.

large tuning-fork *d* vibrating **180** times per second. **A metal** plate *e*, **attached** to one end of the fork, touched alternately **two** wires *h* and *i*, leading respectively to the (+) and (—) poles of a Holtz machine. Two Leyden jars, *r* and *s*, kept the supply of electricity tolerably constant.

In order to prevent an accumulation of either **one** kind of electricity, the wire *g* **was connected to** earth through a long and very fine glass tube, *u v w*, filled with distilled water. This allowed any permanent charge to slowly escape; but, by reason of its very small conducting power, did not perceptibly diminish the effect of any one of the alternating charges in the $\frac{1}{180}$ part of a second during which they lasted.

CASTING THE BALLS.

The balls of sulphur **and** other fusible substances were very carefully cast in a kind of bullet-mould, and very slowly cooled in order that there might **be no cavities.** In order to prevent the balls sticking, the inside of **the mould** was previously rubbed with a little oil, or, in the case **of the** resin balls, wetted with a little distilled water.

The resin balls **were not wiped, but the water** was allowed to evaporate from them.

The sulphur **balls were** carefully cleaned, and then any electrification which **they might** have acquired by friction was discharged by passing them through the flame of a spirit lamp. Sometimes the **balls were discharged by** being hung up for a long time under a bell-glass.

The ebonite balls were turned in a lathe.

Two balls of each substance having been selected of as nearly as possible equal weights, the heaviest was chosen to be the " metal ball," and fine holes were drilled in it until it was exactly of the same weight as the other. It was then gilt by being covered with gold leaf.

CORRECTIONS.[*]

The effect of changes in the potential of the working ball was allowed for by noting the deflections of the arm P P'.

The corrected deflections of the balance were set down as " reduced attractions."

It was found that slow conduction by the silk threads introduced errors into these experiments, and therefore threads of shellac were substituted.

In order to prove that there was no action on the shellac threads themselves, the sulphur ball was hung on the little hook o, fig. 28, where it was protected by the gilt paper box, and the working ball electrified. No deflection of the arm took place.

The fact that the metal and sulphur balls when attracted did not come to rest in the same position, but at different distances from the working ball, was allowed for.

A correction was introduced for small differences in the volumes of the sulphur and metal balls.

It was found necessary to line the greater part of the glass with tinfoil connected to earth, as if this were not done it gradually became charged.

THE ELECTRIFICATION.

In these experiments the surface of the working ball was changed to a potential corresponding to a spark-length of from one to three millims.

The radius of the working ball in the final experiments was $20\frac{1}{2}$ millims. and the distance from centre to centre of the balls was in different experiments 49 and 62 millims.

Now as potential diminishes simply with the distance, the

* Some of these corrections are given in a later paper, read October 8, 1874. Wiener Sitz., Bd. lxx. part ii. p. 307.

H

potential of the equipotential surface **passing through** the **centre** of the suspended ball will be from

$$\frac{20\frac{1}{2}}{49} \text{ to } \frac{20\frac{1}{2}}{62},$$

say $\frac{2}{5}$ to $\frac{1}{3}$, of that of the working ball.

Now according **to De La Rue** (see Chap. XLII.), the potential which would **produce a** spark of one millim. would be that of about 2000 chloride **of silver** cells, and **that which would give a** spark **of** three millims. **would** be that of about 3450 cells.

The potentials to which the suspended **balls were** electrified would therefore be **from that** of 600 **to 800 cells with the** short spark, and **from 1100 to 1400 with the long** one.

No considerable difference of capacity was observed within these limits.

EFFECT OF LONG CHARGING.

When the paraffin and resin-balls were charged for any considerable time, their attractions increased. This is a phenomenon of "Residual charge," or as Faraday calls it "Electric absorption." Boltzmann gives it the **very expressive name** of "Dielectric afterworking."

After the charging had **gone on** for from five to ten minutes, it **was found** that a **paraffin ball** was attracted as strongly as a **metal one, while the attraction** on a **sulphur** one **was not** increased at all.

RESULTS.

The **following** table gives the specific **inductive** capacities **calculated from the** experiments where the charging only lasted **for from** $\frac{1}{300}$ to $\frac{1}{64}$ sec., and which therefore were not complicated **by** the phenomena **of residual charge.**

Side by side with **them are** the numbers obtained by the condenser **method, copied** from the table on page 90.

Dielectric.	Specific Inductive Capacity	
	from Attraction Experiments.	from Condenser Experiments.
Sulphur　.　.	3·90	3·84
Ebonite　.　.	3·48	3·15
Paraffin　.　.	2·32	2·32
Resin　.　.	2·48	2·55

COMPARISON OF RESULTS.

The two following tables give the ratios of the attractions observed with different durations of charge and the apparent specific inductive capacities calculated from them :—

Dielectric.	E = Ratio of Attractions.					
Duration of Charging in Seconds.	$\frac{1}{350}$ to $\frac{1}{5}$	0·9	1·8	22½	45	90
Sulphur	2·033	..	2·125	..	2·110	..
Ebonite	2·211	..	2·064	..	2·094	..
Paraffin	3·209	2·980	2·920	..	1·420	..
Resin .	3·025	2·140	1·927	1·730	1·700	1·650

Dielectric.	Apparent Specific Inductive Capacity.					
Duration of Charging in Seconds.	$\frac{1}{350}$ to $\frac{1}{5}$	0·9	1·8	22½	45	90
Sulphur	3·90	..	3·66	..	3·70	..
Ebonite	3·48	..	3·82	..	3·74	..
Paraffin	2·32	2·51	2·56	..	8·12	..
Resin .	2·48	3·63	4·23	5·11	5·28	5·61

We see from this table that the capacity of sulphur only exhibits irregular variations which we may probably account for by errors of experiment.[*] That of ebonite apparently increases up to a certain maximum, and there remains constant. Paraffin shows a steady increase up to 1·8 seconds, and somewhere between 1·8 and 45 seconds something in its structure appears to give way, and the capacity becomes very great indeed. It is to be regretted that Prof. Boltzmann did not make an experiment on paraffin at 22½ seconds.

For resin the increase seems to be perfectly steady and regular up to the longest charging used in these experiments.

Another set of experiments,[†] in which various special precautions were taken, gave for paraffin the result K = 2·343.

---

[*] Probably the ball was partly crystalline, and its axis was not always in the same direction in different experiments. See next page.

[†] Wiener Sitz., Bd. lxix. part ii. page 812.

The duration of charging is not mentioned, but it probably was from $\frac{1}{300}$ to $\frac{1}{64}$ second.

CRYSTALLINE SULPHUR.

On Oct. 8, 1874,[*] Prof. **Boltzmann** read another paper on the inductive capacities of crystalline sulphur. It is found that the inductive capacity of a **crystal is** different in different directions which **are** related to the **axes** of crystallization. On the theory that induction is a state of strain, this difference **would be just what we** should expect.

We know that a crystal of sulphur, for instance, **has three** principal "**axes** of elasticity;" that is, that **there are three** principal **lines in the crystal** along which the elasticities **are** different.

Now the transmission of strain through any medium is affected **by the** elasticity along the direction of transmission, and therefore the specific inductive capacity of any dielectric will vary according **to the** elasticity of **that** dielectric **along the line of** electric force.

In Prof. Boltzmann's experiments spheres of crystalline sulphur were prepared and hung **up** with their axes at different angles to the line of **force, and then the attractions** measured in the manner which **we have just** described.

The following results were obtained for sulphur, where g m **are the three principal axes** of elasticity :—

$$K_g = 4.773.$$
$$K_m = 3.970.$$
$$K_l = 3.811.[\dagger]$$

EXPERIMENTS OF ROMICH AND FAJDIGA.[‡]

Prof. Boltzmann's experiments have **been continued in his** laboratory by two of his students, Messrs. Romich and **Fajdiga.** Their paper was read on Oct. 8, 1874.

The object of their research was to determine whether the differences of attraction **observed depended** on the dielectric

---

[*] Wiener Sitz., Bd. lxx. part ii. page 342.
[†] The importance of these results will be seen in Part IV.
[‡] Wiener Sitz., Bd. lxx. part ii. p. 367.

properties of the whole ball, or merely on the nature of its surface.

If the former were the case, a ball of sulphur thinly coated with paraffin should act as a ball of sulphur; if the latter, it should act as a ball of paraffin.

A series of experiments were made, of which the following were the results :—

Nature of Ball.		E = Ratio of Attraction.	
Sulphur ball	Not covered . .	$\left.\begin{matrix}2{\cdot}05 \\ 2{\cdot}07\end{matrix}\right\}$ . .	. 2·06
	Coated with . ⎱ paraffin . ⎰	$\left.\begin{matrix}2{\cdot}01 \\ 2{\cdot}12\end{matrix}\right\}$ . .	. 2·06
	Coated with resin . . .	. . . . . .	. 2·51
	Lacquered with shellac varnish.	. . . . . .	. 2·01
Paraffin ball ⎰ Not covered	. . . .		3·320
⎱ Lacquered.	. . . .		3·314

These experiments clearly show that the attraction of the ball is determined by its whole substance, and not by its surface.

The only apparent exception is that of the ball covered with resin, but the authors explain that the resin coating was so thick that it formed an important fraction of the mass of the ball.

EXPERIMENTS OF ROMICH AND NOWAK.[*]

At the same sitting of the Academy another paper was read by Messrs. Romich and Nowak, the latter being also a student in Prof. Boltzmann's laboratory. The chief object of their research was to examine the electric absorption, or " after-working," in different substances.

For this purpose they made two sets of measurements. In one the charge was reversed 64 times in each minute (rather more than once a second), and in the other the working-ball was permanently charged in one way during the experiments.

The following table gives the results. The sulphur-graphite ball was made by putting powdered graphite into melted sulphur.

[*] Wiener Sitz., Bd. lxx. part ii. p. 380.

Dielectric.		Specific Inductive Capacity:	
		Determined by Alternate Charges.	Determined by Permanent Charge.
Glass		. . 7·5 . .	. . 159
Fluor spar . . . . { 1		6·7	7·1
. . . . { 2		7·2	9·9
. . . . { 3		6·7	7·9
Quartz . . . . { 1		4·6	∞
. . . . { 2		4·6	1000
Iceland spar { Perp. to optic axis		. . 7·7 . .	. . 9·9
{ Parallel		. . 7·5 . .	. . 8·5
{ Another ball		. . 8·4 . .	. . 8·6
Selenium { 24 hours after casting . . .		. .10·2 . .	. 151
{		{ ∞	35
{ Later		{ ∞	201
{		{ 59	49
Another ball of { 24 hours after casting .		. . 8·4 . .	. .56
Selenium { Later		. . 116 . .	. . 126
Sulphur graphite		. . 4·0 . .	. . 4·4

When the specific inductive capacity is written down **as ∞** (infinite), or very large, it means that the attractions of the metal and dielectric balls were sensibly equal.[*]

These experiments seem to show that with imperfect insulators the specific inductive capacity increases indefinitely as the charging continues. It is very difficult to say what duration of charging is sufficiently short **to get the true instantaneous capacity.**

With perfect insulators, however, the specific **inductive capacity seems a** tolerably constant quantity.

---

[*] The specific inductive capacity of all good conductors is infinite, as we may see from the formula (2) on page 88. For we remember that the insertion of a plate of thickness n and specific inductive capacity K into an air condenser increases its capacity as if the distance between the plates had altered from m, to

$$m - n + \frac{n}{K}.$$ But if a *metal* plate of thickness n be inserted, the capacity will increase as if the distance between the plates had diminished from m to $m - n$. This will be obvious if we consider the metal plate to **touch one of** the condenser plates.

Thus the capacity at the same time corresponds **to a distance**

$$m - n$$

and to a distance

$$m - n + \frac{n}{K}$$

This **cannot** be true unless $\frac{n}{K} = 0$, that is unless $K = \infty$.

Experiments of Schiller.[*]

In 1874 Schiller published an account of some experiments on Specific Inductive Capacity by a method of "electric oscillations," in which the condenser was only charged for about $\frac{1}{20000}$ of a second. The method will be described at the end of Chapter XXX.

He also made some experiments by a slower method, where the charge in each case lasted about $\frac{1}{25}$ second.

In this method the condenser was charged and discharged through a "galvanometer"[†] some 20 to 25 times per second.

The ratio of the deflections of the galvanometer needle, when air and other dielectrics were in the condenser, gave the ratio of the charges of the condenser in the various cases—that is, of its capacities. The following results were obtained :—

Dielectric.	Specific Inductive Capacity.		Duration of Charging.
	By Slow Method.	By Oscillation Method.	
Ebonite	2·76	2·21	·0000652 sec.
Brown India Rubber . .	2·34	2·12	·0000630
Vulcanized ditto. . .	2·94	2·69	·0000706
Paraffin, quickly cooled, nearly transparent . .	} 1·92 {	1·68	·0000568
Ditto, slowly cooled, white .	} 2·47 {	1·81	·0000858
Another Plate . . .		1·89 {	·0000580 to ·0000898
Semi-opaque Glass "Halb-weises"	} ..	2·96	·0000752
Another Plate . . .	..	3·66	·0000838
"Kalk-weises" Glass . .	4·12	..	..
Clear Plate Glass . .	6·34	{ 5.78 5·88	·0000859 ·0001281

[*] Pogg. Ann. 152, 1874, p. 535.

[†] A galvanometer is an instrument which measures the total quantity of electricity which passes through it in a second. See Chapter XX.

EXPERIMENTS OF SILOW.[*]

In 1875 Silow published the results of some **experiments on the** Specific Inductive Capacities of Liquids.

He used an electrometer **of the shape** shown in plan in fig. 33.

It consisted of a cylindrical glass vessel coated inside with tinfoil.

The tinfoil was not continuous, but formed four quadrants, which were connected in the same way as in the Thomson electrometer. The needle consisted of a light rigid rod, carrying at its ends pieces **of metal curved** so as to be parallel with the sides of the cylinder.

The cylinder could be filled with various liquids.

The needle was kept constantly connected **to earth.**

The deflection with a given battery was first noted when **the** cylinder was empty—that is, **when there was only air between** the needle plates and the cylinder.

The cylinder being filled, the deflection with the same battery increased, owing to the greater facility of transmitting induction possessed by the liquid. From the ratio of the two deflections the specific inductive **capacity of** the liquid could be **calculated.**

If the electrometer had been perfectly symmetrical, the specific inductive capacity of the **liquid would have been simply expressed by the ratio of the deflections;** but as it was impossible **to** avoid certain irregularities in its construction, it **was** graduated by means of a series of observations of the deflections produced with different battery powers.

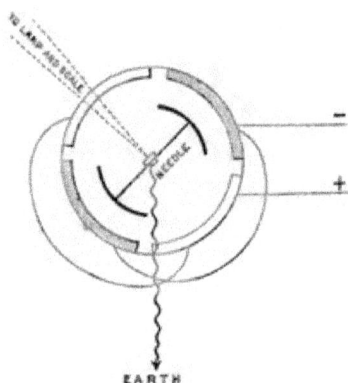

Fig. 33.

The following determinations **were made for oil of turpentine. Two** specimens were tested :—

In different experiments Silow **found for one of them—**

		Specific Inductive Capacity.
Oil of Turpentine I.	. .	2·277
		2·279
		2 2505
	Mean . .	2·2688

[*] Pogg. Ann. 156, 1875, page 389.

For the other he found—

	Specific Inductive Capacity.
Oil of Turpentine II. . .	2·173

SILOW'S SECOND METHOD.

In 1876 the same author published[*] an account of some further experiments made by a different method.

In this method a condenser, battery, and galvanometer are used. By means of a "rapid commutator"[†] the condenser is alternately charged from the battery and discharged through the galvanometer from 900 to 1500 times per minute; that is to say, 900 to 1500 charges of the condenser were passed through the galvanometer in each minute.

Now as the deflection of the galvanometer depends on the total quantity of electricity passing through it in a given time, it will, when the speed is uniform, give the average value of the charges of the condenser. By alternating experiments with air and other dielectrics, the specific inductive capacities of the latter could be calculated.

The following results were obtained :—

Dielectric.		Specific Inductive Capacity.
Oil of Turpentine II.		2·153
Petroleum, 2 specimens { I.	. . . .	2·071
{ II.	. . . .	2·037
Crystallized Benzol		2·198

EXPERIMENTS OF WÜLLNER.[‡]

In 1877 Wüllner published an account of some determinations of the specific inductive capacities of various substances.

He used two methods. In the first method the condenser consisted of two horizontal metal plates. The lower one was connected to earth, and the dielectric laid upon it. The upper plate was suspended horizontally by a cord passing over a pulley so that it could be raised and lowered. The distance between the plates was observed by means of a microscope.

The condenser being charged, the potential was observed; first, while the dielectric was between the glass plates; secondly, after it had been withdrawn.

The ratio of the two capacities of the condenser was then the inverse ratio of the potentials in the two cases.

[*] Pogg. Ann. 158, 1876, p. 306.   [†] Chapter XX.

[‡] Sitzungsb. königl. bayer. Akad., 1877, page 1.

In order to calculate the effect of leakage, the upper plate was drawn up away from the rest of the condenser, both before and after each observation, and the charge measured.

The difference between the **two** measurements was the loss that had occurred during the experiment.

Wüllner's Second Method.

The above method was, however, found to be very inaccurate, so another was adopted. **The** same condenser was used, but a constant potential was obtained by using a battery of 12 cells, one end of which was connected to earth.

The condenser was charged from the battery, and the charge measured by means of a torsion electrometer, and was expressed in terms of the number of degrees of torsion which were required to bring the deflected needle back to a constant zero position.

The following results were obtained :—

Dielectric	Specific Inductive Capacity.
Paraffin	1·96
Ebonite	2·56
Sulphur	2·88—3·21
Shellac $\begin{cases} \text{I.} \\ \text{II.} \end{cases}$	3·73 2·95
Glass	6·10

Wüllner's General Conclusions.

Wüllner arrived at the following general conclusions on the subject :—

"The thickness of **the** dielectric does not affect its specific inductive capacity.

"The capacity increases if the electrification is continued—at first rapidly, then more slowly, and then it gradually approaches a fixed maximum value.

"The capacity which will be reached in **a given time is** increased by frequent repetitions of experiments on the same plate, and also by long charging. By frequent charg**ing**, an increase **of** capacity is produced in sulphur, which lasts for a very long time.

"The change is not **a permanent** one, but, after a long time, it gradually disappears.

"*The instantaneous capacity (that is, the capacity when the charging only lasts for a very small fraction of a second) is of quite a different nature to that capacity which increases slowly as the electrification continues. The 'instantaneous capacity' is independent of the conductibility—the 'slow capacity' is not.*"

If the above conclusion should be confirmed by future experiments, it will probably at once explain the great discrepancies in the results obtained by different observers; for, in experiments of different durations, the capacity observed would be partly due to the instantaneous capacity, and partly due to whatever value the slow capacity might have reached at the conclusion of the experiment. An argument in favour of this view is found in the fact that the results obtained by different observers agree much better for good insulators than for bad ones, for, according to it in a perfect insulator the two capacities should be equal.

DR. HOPKINSON'S EXPERIMENTS ON OPTICAL FLINT GLASS.[*]

On May 17, 1877, a paper by Dr. Hopkinson on the "Electrostatic Capacity of Glass" was communicated to the Royal Society.

The instruments used were:—

(1) The sliding condenser made for Messrs. Gibson and Barclay's experiments; [†]

(2) A Thomson electrometer (White pattern);

(3) A battery of 42 Daniell cells,[‡] the middle of which was connected to earth.

It is sufficient for our present purpose to know that this is an arrangement by which equal and opposite potentials can be obtained simultaneously.

(4) The guard-ring condenser.

This consists of a fixed insulated brass disc, k (figs. 34, 35), surrounded by a flat ring, such that the ring and disc together form a large flat circular plate with an annular space of one millim. round the disc.

Below this is another larger plate, e, which can be moved up and down by a micrometer screw.

The glass plates under examination are placed between e and k so as to form the dielectrics of the guard-ring condenser.

The glass being inserted, the sliding condenser is moved until it and the guard-ring condenser have the same capacity, and the screw of e is read. The glass is then removed, which diminishes the capacity of the guard-ring condenser.

---

[*] *Phil. Trans.*, 1878, page 17.   [†] See page 85.
                [‡] See Chapter XVIII.

e is then screwed up until the **capacity is again equal** to that of the sliding condenser.

The screw is again read, and from the difference of readings **the** difference of capacity caused by the glass can be calculated.

COMPARISON OF THE CONDENSERS.

The equality of the condensers is determined as follows :—We compare the sliding condenser with that consisting of *k* and *e* without the guard-plate.

The connections are first arranged as in fig. 34, and the battery

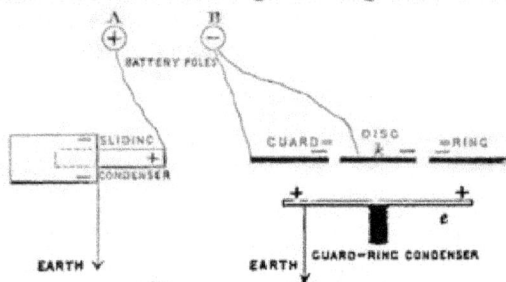

FIG 34.

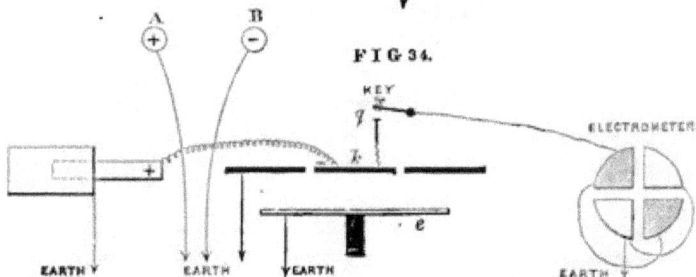

FIG. 35.

then charges the **disc *k* with** its guard-ring, **and the** inside of **the sliding** condenser, **to equal** and opposite potentials.

If the capacities are equal, the charges will also **be** equal and **opposite.   If the** capacities **are** unequal, **the charges will be un-equal.**

By a sudden movement of an ingeniously contrived " switch," **the connections are now altered to those shown in** fig. 35.

The battery, **being put to earth, exercises·no** further effect, **and the (+) charge of the sliding condenser is combined with the (−) one of *k*.**

If the capacities **are equal, the resulting charge will be zero.**

To test this, contact is now immediately made at q (fig. 35), which puts the condensers into connection with the electrometer. If the capacities are equal, there will be no deflection.

If the sliding condenser is the largest, there will be a (+) deflection, and e must be raised; and if the guard-ring condenser is largest, there will be a (—) deflection, and e will have to be lowered.

The only use of the guard-plate is to facilitate calculation, for, as we stated in the account of the absolute electrometer, the distribution of the electricity near the edges of a plate is complicated, but near the centre it is uniform. The effect of the guard-ring is to make the disc k a plate which is all centre.

The following experiments were made : —

Kind of Glass.*	Specific Gravity.	Specific Inductive Capacity.	S.I.C. / Sp. Gr.
Very light Flint . .	2·87	6·57	2·29
Light Flint . . .	3·2	6.85	2·14
Dense Flint . . .	3·66	7·4	2·02
Double extra dense Flint	4·5	10·1	2·25

We see that there seems in these experiments to be some connection between the density and the capacity. It is, however, probably not a very important one.

GORDON'S EXPERIMENTS.†

The great difficulty which all investigators of specific inductive capacity have met with has been due to the fact that, if a dielectric is charged for any appreciable time, some of the charge is " absorbed," and the phenomena of " residual charge " complicate the observations.

The present writer has made some experiments on the specific inductive capacities of various substances by a method where the effects of absorption are guarded against in two ways.

* The glasses were those made by Messrs. Chance for optical purposes.
† *Phil. Trans.*, 1879, page 417.

(1) The electrified metal plates of the condenser do not touch the dielectrics; and

(2) The charging only lasts $\frac{1}{12000}$ of a second.

THE INDUCTION BALANCE.—PLATES VIII., IX., X.

The chief instrument used is a very complicated condenser called the " Induction Balance," the general plan of which is due to Sir Wm. Thomson and Professor Clerk Maxwell.

It consists essentially of five circular parallel metal discs, *a b c d e; b c d e* are fixed, and *a* can be moved parallel to itself by means of a screw. *a c e* are 6 inches in diameter, *b d* are 4 inches. There is a space of about 1 inch between each plate and the one next it.

We will give the details of its construction immediately, but first we will explain the method of its working.

THEORY OF THE INDUCTION BALANCE.—PLATE VIII.

The source of electrification (coil poles, in Plate VIII.) is one that gives equal and opposite potentials. Where the double sign ($\pm \mp$) is given, it means that the sign of the electrifications can be rapidly reversed. We will for the present consider the electrifications of all parts of the apparatus to have the upper sign, and the reversing engine not to be at work.

We see that one pole of the coil is connected to the outside plates *a* and *e*, and the other pole to the centre plate *c*.

The two small plates, *b* and *d*, are connected to the quadrants of an electrometer.

The centre plate *c* can never produce any deflection of the needle, because, being placed half-way between the small plates, it will produce an equal and similar charge on each of them, and therefore equal and similar charges on the quadrants.

The outer plates will also produce no effect on the needle as long as there is only air in the balance and they are placed symmetrically—that is, as long as distance *a b* is equal to distance *e d*.

If, however, *a* is moved by its screw away from *b* (so as to make distance *a b* greater than *e d*), there will be a *less* inductive action from *a* to *b* than from *e* to *d*, and the needle will be deflected in the direction which shows that the unshaded quadrants are most strongly electrified.

But if, on the other hand, any dielectric of greater specific

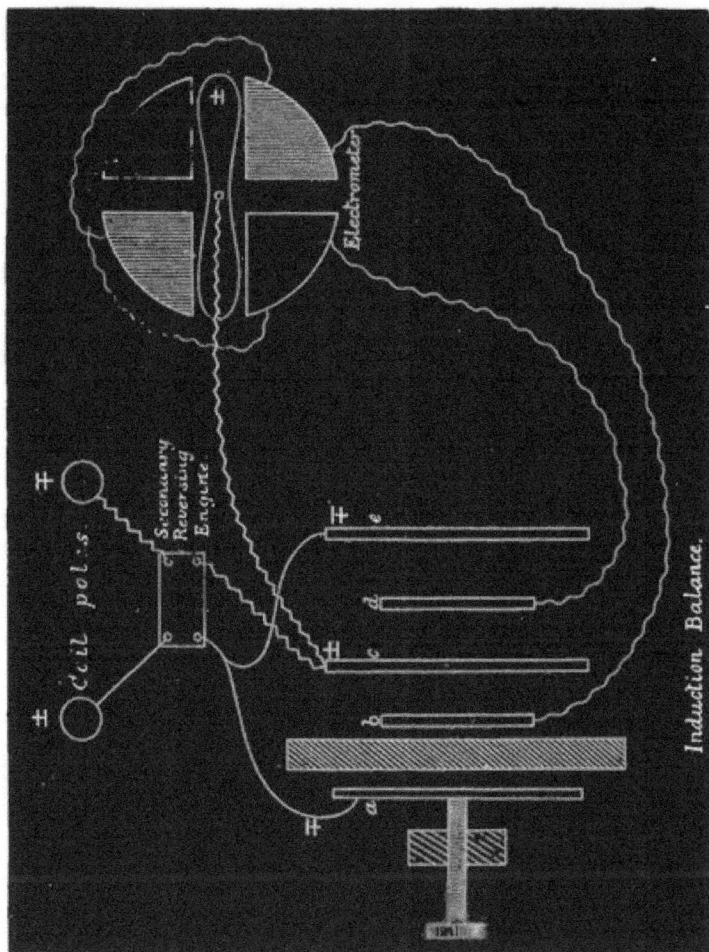

Coil poles.

Secondary
Reversing
Engine

Electrometer

Induction Balance.

PLATE VIII.—DIAGRAM OF INDUCTION BALANCE.

inductive capacity than air be placed between a and b, there will be a *greater* inductive action from a to b than from e to d, and the needle will be deflected in the opposite direction.

It is clear, then, that if we insert a dielectric, and at the same time screw a away from b, we can find a position for a where the increased induction through the dielectric is exactly balanced by the decrease due to the greater distance, and then the needle will remain at zero.

The distance which a will have to be moved to compensate any given dielectric plate will depend only on the thickness of the latter, *and on its specific inductive capacity.*

In the experiments we read the position of a, which brings the electrometer needle to zero : 1st, when there is only air in the balance ; 2nd, when the dielectric is inserted.

The difference of these two readings is the distance which a has been moved. We measure the thickness of the dielectric, and then we can calculate the specific inductive capacity by the following mathematical formula.

THE FORMULA OF CALCULATION.[*]

Let the reading of a, when there is only air in the balance, be a_1, and that when the dielectric is inserted, a_2 ; then (a_2-a_1) is the distance which a has had to be moved.

A dielectric plate of thickness b, and specific inductive capacity K, acts like a plate of air of thickness $\frac{b}{K}$. That is to say, the capacity of a condenser whose dielectric plate had a thickness b, and specific inductive capacity K, would be equal to that of a similar condenser having, for its dielectric, a plate of air of thickness $\frac{b}{K}$.

We must remember that when we insert a dielectric plate of thickness b into the balance, we displace a plate of air of the same thickness b.

The effect then of inserting the dielectric is to *increase* the capacity of the condenser, consisting of the plates a and b, as much as if we had brought the plates nearer by a distance b, and further apart by a distance $\frac{b}{K}$; that is (as b is

* Compare page 87.

greater than $\frac{b}{K}$), as if we diminished the distance between the plates by a quantity $b - \frac{b}{K}$.

But as we have moved a so as to keep the needle at zero, we have produced an exactly equal *decrease* of capacity by increasing the distance between a and b by a quantity $(a_2 - a_1)$. It is clear that this real increase of distance must be exactly equal to the imaginary decrease of distance produced by the dielectric, and we must have

$$b - \frac{b}{K} = (a_2 - a_1) \; ;$$

or, in other words,

$$K = \frac{b}{b - (a_2 - a_1)},$$

and this formula was used to calculate the results of the experiments.

We note that we do not require to know the distances $a_1\ b$ or $a_2\ b$, but only their difference, which is much more easily measured.

The Reversals.

We have hitherto supposed the needle to be charged positively in the ordinary way. Let us suppose the equilibrium not to be established, but the action of a to be greater than that of e, and the electrifications to have the upper signs. a and e will induce ($+$) electricity on b and d, and ($-$) on all four quadrants; but the electrification of the shaded quadrants will be the strongest, and the needle will turn to the right (in the direction of the hands of a watch).

Now, suppose the electrifications of $a\ c\ e$ to be all reversed in sign, but to have the same numerical values as before; a and e will now induce a ($-$) charge on b and d, and a ($+$) one on the quadrants. The shaded quadrants would still be the strongest, but the needle would now turn to the *left*. It is clear that if, as in actual work, the reversals were very rapid, the needle would merely receive rapidly alternating impulses in the two directions, and no disturbance of the equilibrium would produce any deflection.

To escape from this difficulty, Prof. Clerk Maxwell arranged *that the needle, instead of being permanently charged, should be connected to the plate c.*

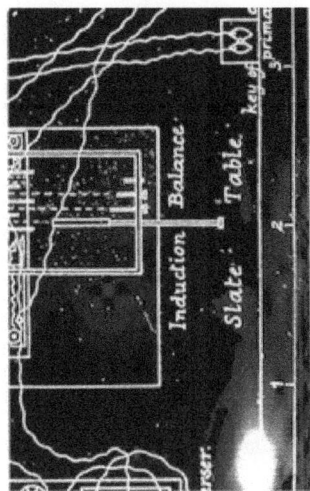

PLATE IX.—PLAN OF INDUCTION BALANCE CONNECTIONS

The sign of the charge of the needle is then reversed with the reversals of the charges of the balance.

Let now the electrifications have the upper signs, and the shaded quadrants be strongest. The quadrants will be (−) and the needle (+), and, the force being attractive, the needle will turn to the right.

Now let the electrification be reversed, the quadrants will be (+) and the needle (−). The force will still be attractive, *and the deflection in the same direction as before.*

In practice, when the electrifications of the five plates, the dielectric, the needle, and the four quadrants were all being reversed 12,000 times per second, the deflection of the needle was perfectly steady, and exactly under the control of the screw of *a*.

A motion of *a* of $\frac{1}{1000}$ inch usually moved the spot of light on the electrometer scale about one millim.

The rapid reversals (12,000 per second) were obtained by means of an induction coil and high-speed break, which are described in Part III.*

The secondary reversing engine, described in Chapter XX., was used to again reverse the electrifications on their way to the balance about 30 times per second, in case there should be any preponderance of either (+) or (−) after the first reversal.

DETAILS OF THE CONNECTIONS.—PLATE IX.

Plate IX. is a ground-plan of the laboratory, showing the arrangement of all the instruments and their connecting wires.

The student who is ignorant of the theory of the induction coil is advised to defer the consideration of this plate until after he has read Chapter XL.

DETAILS OF THE INDUCTION BALANCE.

Plate X., and figs. 36, 37.

Plate X. shows the induction balance and electrometer in perspective.

The electrometer is enclosed in the wooden case at the right of the picture. Its lamp, scale, and lens are seen at the back. The candle at the left illuminates the scale of plate *a*.

* See Chapter XL., and plates in it.

I

Size.

The total height of the balance is 2 ft. 2 in., and the size of the table shown in Plate X. is 4 ft. 6 in. by 3 feet.

The Slate Table.

The whole apparatus stands upon a massive brick and slate table, which is supported on a mass of brickwork independent of the floor of the laboratory. Thus no vibration is communicated to the apparatus by persons moving about.

The Balance.

The plates *b c d e* are supported from above by steel rods. The lower end of each rod is screwed into the upper edge of a plate, the upper into an ebonite plug fixed into a small triangular horizontal brass plate, at the corners of which are levelling

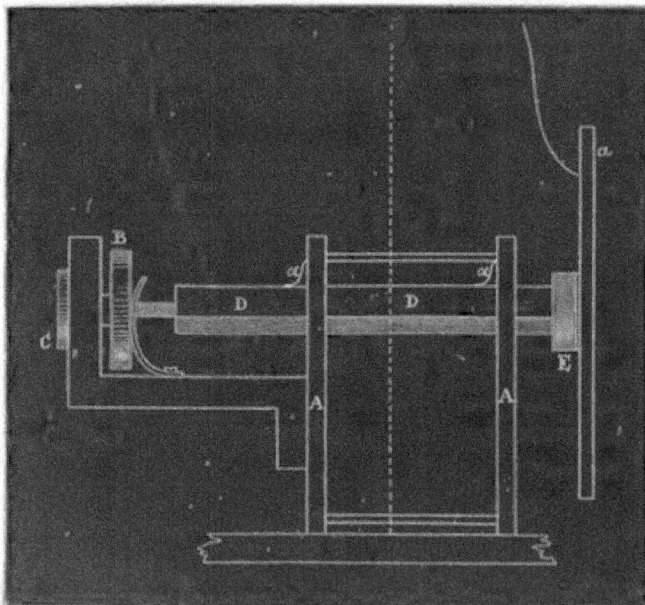

Fig. 36.

screws. The screws rest on a flat brass stage, a slit in which allows the rod to pass through. As there is not room for all four triangles side by side, there are two stages one above the other.

PLATE X.—THE STATIC INDUCTION BALANCE, ON SLATE AND BRICK TABLE.

The plates *c e* hang from the lower stage, and *b d*, which are furnished with longer rods, from the upper one. The triangles, when adjusted, are clamped by screws, which for *c c* are fixed in the second stage, while for *b d* the clamps are carried on a third stage made especially to hold them. Four stout brass pillars support the three stages. The feet of these pillars are screwed into a large brass plate let into the wooden base of the instrument. On this plate also stand the pillars carrying plate *a*.

Plate *a* is fixed to the end of a brass rod of section , from which it is insulated by a block of ebonite (E., fig. 36). This rod slides on two pillars (A. A., figs. 36, 37), which it touches only on its inclined surfaces (fig. 37). It is pressed downward by stout springs (*a a*, figs. 36, 37). It is moved backwards and forwards by a screw B, pressed by a spring against a hardened steel plate at C (fig. 36). The screw inside D turns in two collars—one fixed to D, and the other only kept from revolving, and forced away from the first by a stout spiral spring. This prevents what is called " backlash," *i. e.*, it ensures that the longitudinal motion shall be reversed at the same time as the motion of the screw.

Fig. 37.

A scale divided to $\frac{1}{50}$ inch is engraved upon D, and a vernier fixed to one of the uprights reads the position of plate *a* to $\frac{1}{1000}$ inch. The scale is read by a microscope, fixed some three inches distant on the case of the instrument, and shown at the left of Plate X., just above the candle.

The five plates are enclosed in a glass case, like a balance case, about 15 inches long. It passes below the stages, and holes in the top glass admit the four steel rods. A hole in the side admits the rod D, so that *a* is inside the case, while the screwhead is outside. The dotted line in fig. 36 shows the position of the glass side of the case.

The five plates are placed close to one end of the case. On

the base, inside the case, are slides, in which stages for carrying the dielectrics move. The stages can be moved by rods projecting beyond the case. Thus a dielectric under examination can be placed between two of the plates, or moved from them, without opening the glass case.

On the upper **fixed stages** are ebonite pillars with double binding screws on **them**; to the lower **nut** of each is attached a flexible spiral **wire, leading to one of the plates** by way of the **steel** rod. By means **of the upper nuts the** plates **can be** connected to other instruments.

THE MECHANICAL SLIDE.

One of the sliding stages—viz.: that **used to place a** dielectric **between** a **and** b *—has " mechanical motions." In** addition to **the rod by which it** is drawn in and out, there are three **other** parallel rods, with milled heads. Turning one gives a lateral motion—viz.: moves any dielectric placed on the slide nearer a or b. By turning the second, **the** dielectric plate can be placed either exactly vertical or inclined a little in either direction; and turning **the** third gives **the** dielectric a small angular motion round a vertical axis.

THE BRASS COVER.

The upper **stage, which carries the connections of the small** plates b d, is enclosed in **a brass box, shown** in outline in Plate X. The steel rods belonging to the plates b d **are enclosed in** metal tubes.

The wires leading to the electrometer from **the** terminals of the small plates on the upper stage are also enclosed in a wide brass tube. The electrometer itself is enclosed in a wooden box, lined with tinfoil; a small window (not shown) allows the light of the lamp to enter and return. The box, tube, and tinfoil are carefully connected to earth. The tube passes through the base of the balance to allow the glass shade (not shown), with which it is covered when not in use, to be lifted on and off without disturbing the connections.

OTHER DETAILS.

The wire which passes just above the electrometer leads from plate c to the needle.

* This is the only stage that is used in actual work.

The little plug, seen in the front of the picture, makes and breaks the coil primary (see Chapter XLIII.)—that is, starts and stops the electrification.

The observer, standing where he can see the scale of the electrometer, is able to start and stop the electrification with his right hand, and to work the screw of a with his left.

A glass plate is seen on the slide ready to be pushed in between a and b.

The Thomson Electrometer.

This is a quadrant of the simple form by Elliott; one of White's, which I have, was found to be unsuitable for use with the reversing gear. When, as in this case, the instrument is only used as an electroscope, the superior sensitiveness of Elliott's pattern gives it great advantages.

The Callipers.

These are a pair with especially long jaws, made for use in adjusting the plates of the balance. When laid upon a bracket fixed to the outside of the balance case, the jaws projected right in between the plates. This bracket could be inclined either up or down, for measuring at the lower or upper parts of the plates. The "outsides" scale was used in the ordinary way for measuring the thickness of the dielectric plates.

The Dielectrics.

The dielectric plates were 7 inches square, and from $\frac{1}{4}$ inch to 1 inch thick. They were made with their sides accurately parallel. When practicable, they were cut in a planing machine.

The Electromotive Force.

The electromotive force used was about equal to that of 2050 cells of Mr. De La Rue's battery. (Chapter XVIII.)

Specimen Observations.

The following are the details of the determination of the capacity of Light Flint Glass :—

$$b = \cdot 699.$$

a_2	(1) 1·503	(3) 1·498	(5) 1·495	(7) 1·500
a_1	(2) 1·035	(4) 1·034	(6) 1·032	(8) 1·032
$a_2 - a_1$	·468	·464	·467	·468

Mean ·46675.

$$K = \frac{\cdot 699}{\cdot 699 - \cdot 46675} = 3 \cdot 013$$

RESULTS.

The following is a general table of the results obtained for different substances :—

Name of Dielectric.	Specific Inductive Capacity.
Glass—	
Double extra-dense Flint . .	3·164.
Extra-dense Flint . . .	3·054.
Dense Flint	..
Light Flint	3·013.
Hard Crown	3·108.
Common Plate, two specimens—	
No. 1	3·258 } mean 3·243.
„ 2	3·228
Ebonite, four slabs—	
No. 1	2·2697
„ 2	2·2482 }mean 2·284.
„ 3	2·3097
„ 4	2·3077
Best quality Gutta Percha .	2·462.
CHATTERTON's compound .	2·547.
India-rubber—	
Black	2·220.
Grey vulcanized . .	2·497.
Solid Paraffin, specific gravity at 11° C. = ·9109, melting point 68° C. Six slabs cut in planing machine; results corrected for cavities—	
No. 1	1·9940
„ 2	1·9784
„ 3	1·9969 } 1·9936.
„ 4	2·0126
„ 5	1·9654
„ 6	2·0143
Shellac	2·74.
Sulphur	2·58.
Bisulphide of Carbon . .	1·81.

EBONITE.

The agreement of the experiments on ebonite is a very good test of the accuracy of the method and of the formula, for the four plates of it were of very different thicknesses, as the following table shows :—

SUMMARY OF EXPERIMENTS ON EBONITE.

Slab.	Thickness.	Specific Inductive Capacity.
	inch.	
No. 1	·754	2·2697
,, 2	·509	2·2482
,, 3	·516	2·3097
,, 4	·264	2·3077
Mean		2·284

Extreme difference from the mean $= -1·5$ per cent.

CAVITIES IN PARAFFIN.

It is impossible to obtain thick plates of paraffin free from numerous cavities, so a correction had to be applied for them. The best correction which I could devise was, in the calculations, to substitute for the plate of thickness b an imaginary plate of thickness b' where b' is the thickness that the plate would have had if it had been of the same length and breadth, contained the same quantity of paraffin, and had had no cavities.

To determine b' we have—

$$b' = b \frac{\text{Specific gravity of plate}}{\text{Specific gravity of paraffin}}.$$

Let us call the ratio of the specific gravities ϕ and write

$$b' = b\phi.$$

The specific gravity of a small piece of the paraffin free from cavities was determined in the ordinary way and found to be ·9109 at 11° C.

The specific gravity of each plate was determined by weighing in water.

The following table gives the results of the experiments and the correction applied to each plate.

K is the capacity calculated from the real thickness b, and K that calculated from the corrected thickness b'.

SUMMARY OF EXPERIMENTS ON PARAFFIN.

Plate.	Specific Gravity at 11° C.	φ.	b.	b'.	K.	K'.
No. 1	·8783	·9642	·730	·7038	1·9261	1·9940
„ 2	·8771	·9628	·750	·7221	1·9084	1·9784
„ 3	·8951	·9826	·748	·7249	1·9358	1·9969
„ 4	·8909	·9780	·782	·7648	1·9697	2·0126
„ 5	·8933	·9868	·755	·7450	1·9408	1·9654
„ 6	·9021	·9903	·754	·7467	1·9947	2·0143

$$\text{Mean value of} \begin{cases} K = 1·9459 \\ K' = 1·9936 \end{cases}$$

Among the corrected values the extreme difference from the mean is $-1·4$ per cent.

Among the uncorrected values the extreme difference from the mean is $+2·5$ per cent., which shows that, though the correction is a rough one, it gives numbers more nearly accordant with each other than the uncorrected experiments.

The melting point of the paraffin was 68° C.

BISULPHIDE OF CARBON.

The liquid was contained in a glass trough. In this case a_1 was the reading when the empty trough was placed in $a\,b$; a_2 the reading when it was full. The trough was 7 inches wide and 9 inches high; b is its **internal** thickness.

SECULAR **CHANGES** IN THE SPECIFIC INDUCTIVE CAPACITY OF GLASS.*

The experiments on optical glass described above were made about Christmas, 1877, when the glass had only been cast a few weeks.

At the end of July, 1879, I commenced a repetition of the experiments, using the same slabs of glass, and was surprised to find *a large increase in the specific inductive capacity in every case.* In some cases the increase was as much as twenty per cent.

The following is a table of the results :—

* Report of the British Association, Sheffield, 1879, page 250.

SPECIFIC INDUCTIVE CAPACITY OF OPTICAL GLASS.

	Christmas, 1877.	July and August, 1879.
Double Extra Dense Flint .	3·164	3·838
Extra Dense Flint . . .	3·053	3·621
Light Flint	3·013	3·443
Hard Crown	3·108	3·310

The arrangement of the apparatus (including the coil and rapid break) was precisely the same as in my earlier experiments. The electromotive force was as nearly as possible the same, and experiment has shown that moderate variations in it do not affect the results.

The differences observed might have been caused by any one of three things ;—

(1.) By error in the 1879 experiments ;

(2.) By error in the 1877 experiments ;

(3.) By a change in the specific inductive capacity of the glass between Christmas, 1877, and July, 1879.

Careful repetition of the 1879 experiments has convinced me that there is no error in them.

If the difference is caused by error in the 1877 experiments, then in 1877 I must have obtained too low a result. With my induction balance, too low a result is given if the dielectric is covered with a badly-conducting film ; * the effect of covering the dielectric with a well-conducting film is to prevent observation.

Before rejecting the second explanation of the difference, based on the hypothesis of error in the 1877 experiments, it is therefore necessary to prove that in 1877 there was no film on the surface of the glass of sufficient conducting power to cause a large error in the results.

In 1877 the glasses were not washed by immersion in water, but were thoroughly cleaned with a glass-cloth and wash-leather. To the best of my recollection they were first rubbed with a damp

* For the film must be considered to be connected to earth.

cloth, then with a dry one, and then polished with the leather, being frequently breathed on during the process, and then usually warmed at the fire. This process was so far efficacious in removing any conducting film of moisture from the glasses, that at the end of it they were usually found to be electrified by the friction of the leather. When this occurred they were passed rapidly a few times over the **flame** of a spirit-lamp to discharge them. They were always so warm that any visible moisture deposited by the spirit-lamp disappeared instantly.

In the 1879 experiments, which are quoted **in** the preceding table, the glasses **were** washed in hot water, **wiped** and polished, and passed over the spirit-lamp while **still hot.** After observing a difference in the first two specimens examined, I **made preliminary** experiments on the **other two** before cleaning **them.** The following are the results obtained :—

Light Flint Glass.

	Specific Inductive Capacity.
Christmas, 1877	3·01
August 4, 1879.—Dusted lightly with duster, not rubbed	2·90
August 4.—Cleaned in hot water, experimented on while hot	3·44
August 4.—Cooled under tap, wiped with glass-cloth	3·44
August 5.—Had stood twenty-four hours uncovered on table, not wiped	3·39
August 5.—Smeared all over with oil	3·48
August 4.—Smoked on oily surface over paraffin lamp, so as to make glass semi-opaque	3·46
August 5.—Glass made very wet with solution of sal-ammoniac	Experiment impossible.
August 5.—Roughly dried with duster ; surface appeared opaque, like ground glass	1·64
August 5.—Wiped over with glass-cloth, but not rubbed	2·36
August 5.—Rinsed under cold tap, and wiped with glass-cloth, but not polished	3·46
August 5.—While still cold, passed over spirit-lamp till much more clouded than ever would be the case in actual work ; placed in balance, and experiment made as quickly as possible	3·49

Hard Crown Glass.

	Specific Inductive Capacity.
Christmas, 1877	3·108
August 7, 1879.—Not wiped for more than a year ; placed in balance covered with dust exactly as taken from box, which does not shut air tight	3·236
August 8.—Cleaned in hot water, as described above	3·310

My conclusion from the above numbers is that, although it is possible, by sufficiently wetting the surface of the plate, to produce *any* apparent reduction of the specific inductive capacity, yet that, if even very much less care had been taken to clean the plates than was taken in 1877, the greatest quantity of moisture that could accidentally have been left on them would have been totally incapable of producing anything like the difference now under examination.

I am therefore led to the conclusion that in the course of a year and a half an actual change has taken place in the glasses, which is shown by a considerable real increase in their specific inductive capacities. To complete our knowledge of this new phenomenon we require a series of monthly observations, extended over perhaps a period of several years.

I am hoping shortly to continue the investigation of this subject.

All that we can say at present is that the theory of specific inductive capacity is extremely obscure, and that induction, so far from being a "direct action at a distance," is most certainly transmitted by the particles of the dielectrics, and is affected by almost every molecular change which may occur in them.

I am hoping also to investigate the effect of duration of electrification, temperature, and electro-motive force. (See page 141.)

Specific Inductive Capacity of Gases.
Experiments of Boltzmann.[*]

On April 23, 1874, Professor Boltzmann read a paper before the Vienna Academy, in which he announced that he had succeeded in detecting and measuring differences in the specific inductive capacities of gases.

The Condenser.

The condenser (fig. 38) consisted essentially of two metal plates, *d* and *e*, insulated from each other and contained in a glass receiver, which could be connected to an air-pump and to a gas-holder. The plates *d* and *e* could be charged or connected to the earth or to the electrometer by means of two wires, *p* and *q*, which passed through shellac plugs inserted in the top of the receiver.

---

[*] Wiener Sitzungsber., Bd. lxix. part ii. page 795.

The plates *a*, *b*, *f*, *g* had no electrical use, but were kept always connected to earth. Their object was to conduct away any heat which might be developed by the friction of the entering gas.

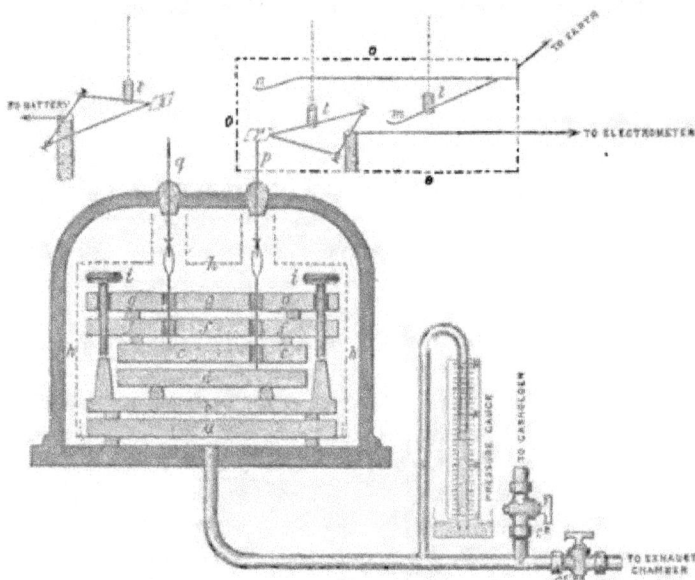

Fig. 33.

The plate *d* rested on three shellac blocks, which were fitted into brass sockets fixed to the plate *b*.

The plate *e* was attached by shellac blocks to the plate *f*, and the latter was levelled by means of three screws, *i i i*. A case, *h h*, of thin sheet brass protected the condenser from external induction. The wires *p* and *q* had each a flexible joint in them, in order that vibrations of the receiver might not shake the condenser plates.

THEORY OF THE EXPERIMENTS.

Before commencing the experiments the plate *e* was connected to earth and the plate *d* to the electrometer, and air pumped in and out of the receiver. No deflection occurred, which showed that the plates were not electrified by the friction of the air.

At the commencement of an experiment, *d* was connected to earth and to the electrometer, while *e* was electrified (+) by means of 300 Daniell cells. The (+) electricity of d escaped to earth, while the (−) was held by the attraction of the (+) on e.

e was now insulated by raising the wire m, but as this caused no change in the electrical state of the system, there was still no deflection of the electrometer.

If now one more cell were added to the battery, the electric action from e to d would be increased, and there would be a deflection β of the electrometer. It was found that, for one added cell, β was about 60 scale divisions.

But if, without altering the battery, the action from e to d were altered by altering the dielectric between them, there would also be a deflection whose amount would depend on the change of specific inductive capacity, and whose direction would depend on whether that change was an increase or a decrease.

It was found that pumping out the air until the pressure was reduced from that of the atmosphere (760 millims.) to about 10 millims. of mercury caused a diminution of specific inductive capacity which produced a negative deflection (−a) equal to about 8 scale divisions.

The Formula of Calculation.

Let the potential due to one Daniell's cell be p, then, if there are 300 cells, the potential of the whole battery will be $300\,p$, or, generally, the potential due to n cells will be $n\,p$. β is the deflection produced by the addition of one cell of potential p, and hence β is proportional to p.

Let K_1 be the specific inductive capacity of the air in the condenser at the beginning of the experiment, and K_2 that of the gas or rarefied air in it at the end.

Then, by the known theory of a condenser, the potential of every part of it will at the end of the experiment be $\dfrac{K_1}{K_2}$ times that which it was at the beginning. As d and the metal case are connected to earth, their potential is zero at the beginning of the experiment, and $\dfrac{K_1}{K_2}$ times zero, that is zero also, at the end.

The potential of the plate e, which was formerly $n\,p$, becomes $\dfrac{K_1}{K_2}\,n\,p$—that is, it *changes* by a quantity

$$n\,p\,\frac{K_1}{K_2} - n\,p, \text{ or } n\,p\left(\frac{K_1}{K_2} - 1\right).$$

This change in the potential of e causes a deflection of a scale divisions.

But, when one cell is added, there is a change equal to p in the potential of e, causing a deflection of β scale divisions.

Hence, by simple proportion—

$$\text{As } n\,p\left(\frac{K_1}{K_2}-1\right) \text{ is to } a,$$

$$\text{so is } p \qquad\qquad \text{to } \beta,$$

or

$$\frac{n\,p\left(\frac{K_1}{K_2}-1\right)}{p}=\frac{a}{\beta},$$

i.e

$$n\left(\frac{K_1}{K}-1\right)=\frac{a}{\beta},$$

or

$$\frac{K_1}{K_2}=1+\frac{a}{n\beta}\quad\cdots\quad(1).$$

EFFECT OF THE DENSITY OF THE GAS.

Professor Boltzmann has found that, with any given gas, each change in its density causes an approximately proportional change in the quantity by which the specific inductive capacity differs from unity. For instance, if the capacity of a given gas at ordinary pressure were 1·002, then at half pressure it would be 1·001.

Let the specific inductive capacity of a vacuum be taken as unity, and that of any gas at ordinary pressure be $1+\lambda$, then we know that, as the capacities of all gases differ very little from unity, λ will be a very small quantity, and will have either a $(+)$ or $(-)$ sign according as the capacity is greater or less than that of a vacuum.

Let K_1 be the capacity of the gas at a pressure of b_1 millims., and K_2 the capacity at a pressure of b_2 millims.; then

$$K_1 \text{ is proportional to } 1+\frac{\lambda b_1}{760}$$

and

$$K_2 \quad\text{,,}\quad\quad\text{,,}\quad 1+\frac{\lambda b_2}{760}$$

or

$$\frac{K_1}{K_2}=\frac{760+\lambda b_1}{760+\lambda b}$$

Now, as λ is a very small quantity, we may write this

$$\frac{K_1}{K_2} = \frac{760 + \lambda b_1 - \lambda b_2}{760} = 1 + \frac{\lambda\,(b_1 - b_2)}{760}.\ast$$

By combining this with the formula (1), page 126, we have—

$$1 + \frac{a}{\beta n} = 1 + \frac{\lambda\,(b_1 - b_2)}{760}$$

which gives

$$\lambda = \frac{a \cdot 760}{\beta n\,(b_1 - b_2)}\quad \cdot\ \cdot\ \cdot\ (2)$$

and this formula was used in the experiments.†

And we have for any gas—

$$K_{760} = 1 + \lambda,$$

where K_{760} is the specific inductive capacity of that gas at 760 millims. pressure when the specific inductive capacity of a "vacuum" of about 5 to 10 millims. pressure is taken as unity.

EXPERIMENTAL DETAILS.

The connections with the wires q and p were made by metal plates r and s fixed to the end of wire triangles hinged, as shown in fig. 38. They could be raised and lowered by means of cords. For better insulation the cords were not attached direct to the wires, but to little shellac cylinders $t\,t\,t$. By raising r until it touched the wire n, the electrometer was put to earth; or, by lowering m while r rested on p, the electrometer and the plate d were put to earth together. A metal case $o\,o$ connected to earth protected all the electrometer connections from accidental induction.

In order that the exhaustion of the air might be done quickly,

---

* For suppose $\lambda b_1 = 2$ and $\lambda b_2 = 1$, then

$$\frac{K_1}{K_2} = \frac{762}{761} = 1 + \frac{1}{761}$$

and $1 + \dfrac{\lambda b_1 - \lambda b_2}{760} = 1 + \dfrac{1}{760}$, which gives a sufficiently close approximation for experimental purposes.

† The portion $\dfrac{a}{b_1 - b_2}$ of the expression (2) was the mean of a number of experiments with different values of b_1 and b_2.

a large chamber was placed between the air-pump and the tap C_1. The chamber was first exhausted by means of the pump, and then the tap C_1 suddenly opened. The condenser could be filled with any gas by first exhausting it and then opening the tap C_2, leading to the gas-holder.

The thickness of each of the plates was from 4 to 5 millims., and their diameters about 160 millims. The plates d and e were 1 millim. apart.

INSULATING POWER OF AIR.

Some preliminary experiments were made to see whether the air between the plates could be trusted to insulate perfectly, particularly at low pressures. The plate e was charged, and d was put to earth. Then e and d were both insulated for a certain time t, and then d connected to the electrometer. Any deflection which occurred would be due to the leakage of electricity from e to d.

When the time t was 5 minutes, which is much longer than the duration of any experiment, no deflection occurred.

When t was 14 hours, there was a deflection showing that $\frac{1}{1200}$ part of the electricity had leaked over.

When the pressure was diminished to about 3 millims., the electricity passed by disruptive discharge* from one plate to the other; but at pressures between 3 and 760 millims. the insulation may be considered to be perfect.

MANIPULATION OF THE EXPERIMENTS.

The operations were performed in the following order:—

(1) The plates r and s and the wire m were lowered. (2) r and m were lifted. (3) The gas was pumped out. (4) r was lowered, but not m, and the deflection of the electrometer was written down as $-a$.

The order of the operations (3) and (4) could be reversed without causing any theoretical difference.

RESULTS.

The following results were obtained:—

Temperature 15° — 17° C.

* See Chapter XLIV.

Vacuum (taken as) 1·000000.	
Gases at 760 millims. and 15° – 17° C.	$K_{760} = 1 + \lambda$
Air	1·000558
Carbonic Acid . . .	1·000892
Hydrogen . . .	1·000250
Carbonic Oxide . . .	1·000650
Nitrous Gas (NO) . .	1·000938
Olefiant Gas . . .	{ 1·001208 / 1·001266
Marsh Gas	1·000890

From these results Professor Boltzmann calculates what the values of K_{760} would have been at 0° C.

He obtains—

Vacuum (taken as) 1·000000.	
Gases at 760 millims. and 0° C.	$K_{760 \cdot 0°}$
Air	1·000590
Carbonic Acid . . .	1·000946
Hydrogen . . .	1·000264
Carbonic Oxide . . .	1·000690
Nitrous Gas (NO) . .	1·000984
Olefiant Gas . . .	1·001312
Marsh Gas . . .	1·000944

For convenience of comparison with the results given by other experimenters, I have calculated from the above tables the specific inductive capacities of the various gases when that of air at 760 millims. and 0° C. is taken as the standard, and called unity.

The formula which I have used is as follows :—

We have

$$\lambda \text{ for air at } 0° \text{ C. and } 760 \text{ millims.} = \cdot 000590.$$

Let the specific inductive capacity of any gas, when that of air at 0° C. and 760 millims. is taken as unity, be $1 + \lambda'$; then, if we write

$$\lambda' = \lambda - \cdot 000590,$$

K

we shall get results which will be true within the limits of error of these experiments.*

We have

Gas.	Specific Inductive Capacity.	
	At 0°C. and 760 millims.	At 15°—17° C. and 760 millims.
Air at 0°C. and 760 millims. taken as 1·000000.		
Vacuum ·999410		
Carbonic Acid . . .	1·000356	1·000302
Hydrogen . . .	·999674	·999660
Carbonic Oxide . .	1·000100	1·000060
Nitrous Gas (NO) . .	1·000394	1·000348
Olefiant Gas . . .	1·000722	{ 1·000618 { 1·000576
Marsh Gas . . .	1·000354	1·000300

EXPERIMENTS OF AYRTON AND PERRY.†

In 1877, Professors Ayrton and Perry announced that they had been able to detect and measure differences in the specific inductive capacities of certain gases, they being at that time quite unacquainted with the previous investigation by Prof. Boltzmann on the same subject.

Fig. 39.

Two condensers—" the open condenser " and " the closed condenser "—were used.

* The true value of $1 + \lambda'$ would be got from the proportion—

As $1 + (\lambda$ for air) : $1 + (\lambda$ for gas) :: $1 : 1 + \lambda$.

Thus the true value of $1 + \lambda'$ for carbonic acid would be

$$1 + \lambda' = \frac{1·000946}{1·000590} = 1·0003558.$$

The value obtained by putting

$$\lambda' = (\lambda - ·000590) \text{ is } 1·0003560,$$

or this formula only introduces an error, 2 parts in 3500 in the value of λ, or 2 parts in ten millions in the value of K.

† " On the Specific Inductive Capacity of Gases." Paper read before the Asiatic Society of Japan, April 18, 1877.

The open condenser (fig. 39) consisted of a metal plate, *z*, 1815 square centimetres in area, laid on a stone pillar, *y*, and another, *w*, supported just above it by means of three ebonite levelling screws. The upper plate was prevented from bending by means of small " girders."

The closed condenser (figs. 40 and 41) consisted of a box in which there were 11 plates, each of 324 square centimetres in area, fixed parallel to each other. Nos. 1, 3, 5, 7, 9, 11 were connected to the box, and Nos. 2, 4, 6, 8, 10 to a wire which passed out through a long glass tube, MP.

The tube RR went to the mercurial air-pump.

The ratio of the capacities of the open and closed condenser was

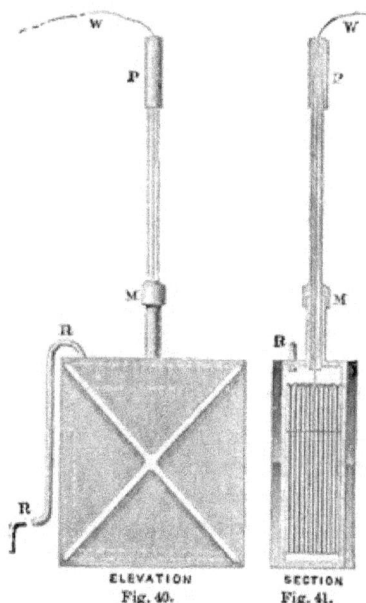

ELEVATION
Fig. 40.

SECTION
Fig. 41.

determined—first, when the latter was full of air at ordinary temperature and pressure, and then when it was filled with some other gas.

The ratio of these two determinations gave the specific inductive capacity of the gas.

The method of comparison was as follows :—

Method of Comparison.

One conductor of each condenser was connected to earth, and the other plates were charged to equal and opposite potentials by means of a battery of 87 Daniell cells and a reversing key.

The key was so arranged that when the (+) pole was connected to one of the condensers, the (−) was connected to earth, and *vice versâ*. The first position of the key gave connections, as shown in fig. 42, and the second as if, in the same figure, the lever *b* had been depressed and *a* raised.

This arrangement gave double the difference of potential which would have been obtained by connecting the two poles to the two condensers respectively. For, if we call the difference of potential

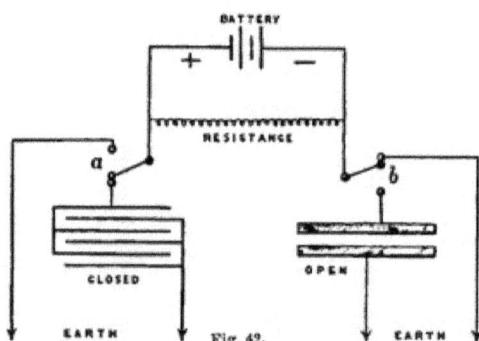

Fig. 42.

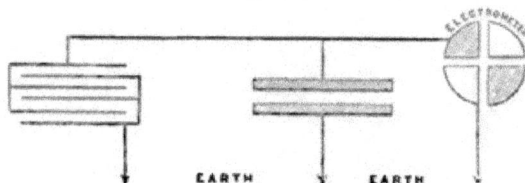

Fig. 43.

between the poles of one cell unity, the latter plan would have caused a difference of potential of 87 between the two condensers.

In the plan which was adopted, one condenser was charged to a potential of 87 less than that of the earth (viz. to − 87), and the other to a potential of 87 more than that of the earth, and their difference of potential was 174.

The two condensers were then disconnected from the battery and connected to the electrometer, as in fig. 43. When the capacities were exactly equal, there was no deflection. When they were not equal, there was a deflection whose direction showed which condenser had the greatest capacity, and whose amount depended on the amount of their difference.

This deflection shows the potential of the remanent charge *in arbitrary units of the electrometer* ; let us call it *p*.

In order to calculate the capacities, it is also necessary to know the potential of the whole charge in the same arbitrary units of the electrometer.

With a small battery this (the potential of the whole charge) could be determined by connecting the poles to the electrometer and observing the deflection. With 87 cells this cannot be done, as the deflection would be greater than could be measured in the instrument.

The following method was therefore employed. The two poles of the battery were connected, as in fig. 42, by a long fine wire, called a "resistance coil of 10,000 units of resistance." [*]

It is known by Ohm's law [†] that the potential falls uniformly from one end to the other of any resistance wire which connects two poles of a battery; and therefore we can obtain a known fraction of it as follows :—

Let us suppose the wire to be 100 inches long. Then the difference of potential between one end and a point one inch from that end will be exactly $\frac{1}{100}$ of the difference between the two ends.

By connecting the electrometer poles to one of the battery poles and to a point, such that the resistance between the electrometer poles is $\frac{1}{100}$ part of the whole resistance between the battery poles, we shall get a deflection which corresponds to $\frac{1}{100}$ of the difference of potential between the latter.

Let us call this deflection *d ;* then the difference of potential between the two poles of the battery will be equal to (100 *d*) arbitrary units of the electrometer.

We know that the charge of a condenser is equal to the product of its potential into its capacity.

Let us call the capacities of the closed and open condensers C and O respectively.

Their charges will then be—

$$(100 \, d. \, \text{C}) \text{ and } (- 100 \, d. \, \text{O}).$$

---

[*] See Chapters XXII. and XXVII.   [†] See Chapter XXI.

When they are added together, they charge the two condensers together to a potential p.

But the capacity will be $C + O$, and therefore the charge will be $p (C + O)$.

Now this is the algebraical sum of the two former charges, and we have the following experimental equation :—

$$100\, d\, C - 100\, d\, O = p\, (C + O)$$

or

$$(100\, d - p)\, C = (100\, d + p)\, O$$

or

$$\frac{C}{O} = \frac{100\, d + p}{100\, d - p}.$$

This determination of the value of $\frac{C}{O}$ having been made with the closed condenser full of air, another was made with it full of some other gas, and the result written down as $\frac{C'}{O}$.

We then have

$$\frac{\frac{C'}{O}}{\frac{C}{O}} = \frac{C'}{C} = \text{Specific Inductive Capacity of the gas under examination.}$$

The following results were obtained :—

Dielectric.	Specific Inductive Capacity.
Air	1·0000
Vacuum	·9985
Carbonic dioxide . . .	1·0008
Hydrogen	·9998
Coal gas	1·0004
Sulphuric dioxide . . .	1·0037

We see that Faraday's failure to detect any difference in the specific inductive capacities of gases was solely due to the extreme minuteness of the differences, and to the want of delicacy of his apparatus.

The " probable error," in any of the above determinations, is calculated to be not more than 0·00015 ; that is, the last figure of each decimal is correct to within $1\frac{1}{2}$.

Summary.

The following table gives a general summary of the results of the experiments described in this Chapter. Their special importance with regard to the theory of electricity will be seen in Part IV.

APPENDIX TO CHAPTER XI.

Professor Boltzmann has had the kindness to send me the following explanation of his formula, given on page 92, for calculating the specific inductive capacity of a dielectric ball from the ratio of the attraction of it, and of a conducting ball, by a charged ball :—

In every dielectric we may write—

$$\mu = k\,R \quad . \quad . \quad . \quad . \quad (1)$$

where μ is the " dielectric moment " of a unit of volume, R the resultant force at any point of the dielectric due to all the electricity acting, whether induced or otherwise, and k is a constant depending on the nature of the body. It is first necessary to prove that

$$K = 1 + 4\,\pi\,k,$$

where K is the specific inductive capacity.

Let us consider one of Faraday's spherical condensers (fig. 25, p. 81). Let the radii of the two concentric balls be

$$a \text{ and } a + \delta$$

respectively, and let there be only air in the space between them.

Let the inner ball be charged to the potential p, and the outer connected to earth. Let the charge of the inner ball be $+$ E and that of the outer $-$ E′.

The potential at a point on the outer surface of the outer ball will be the same as if both charges were collected at the centre, and will be

$$\frac{E}{a + \delta} - \frac{E'}{a + \delta}. \quad . \quad . \quad . \quad (2).$$

But as the outer ball is connected to earth its potential is zero, which gives us

$$E = E' \quad . \quad . \quad . \quad . \quad (3).$$

The potential at any point of the surface of the inner ball due to the charge E′ is

$$-\frac{E'}{a + \delta} = -\frac{E}{a + \delta}.$$

The potential at the same point due to the charge E is $\dfrac{E}{a}$.

The total potential p at the surface of the inner ball is then

$$p = \frac{E}{a} - \frac{E}{a+\delta} = \frac{E\delta}{a^2} \quad \ldots \ldots \quad (4).$$

when δ is very small.

      Now c, the *capacity* of the **condenser**, equals $\dfrac{E}{p}$

(page 67).

We have then

$$c = \frac{E}{p} = \frac{E}{\frac{E\delta}{a^2}} = \frac{a^2}{\delta} \quad \ldots \ldots \quad (5).$$

    Now let the condenser be filled with some other dielectric instead of air. We may consider the dielectric to be divided into a great number of indefinitely thin concentric shells, each of thickness d.

    Every one will be polarized, so that there will be a charge $-e$ on its inside **surface and** $+e$ on its outside. Each $+e$, except that on the outside shell, **will have** a $-e$ next it, and each $-e$, except the one on the inside shell, **will** have a $+e$ next it; so all the charges will neutralize each other, except that there will remain a $-e$ on the inner surface of the dielectric and a $+e$ on **the** outer surface. There are now four charges acting, viz. $+E$, $-e$, $+e$, **and** $-E'$. **The** potential due to all four together at a point on the outer metal sphere will be

$$\frac{E}{a+\delta} - \frac{e}{a+\delta} + \frac{e}{a+\delta} - \frac{E'}{a+\delta} \quad \ldots \ldots \quad (6);$$

and as the outer sphere is connected to earth, this must be zero, which shows that E still equals E′ after the **insertion of the dielectric.**

    The *potential* due to the four charges at a point on the inner ball will be

$$p' = \frac{E}{a} - \frac{e}{a} + \frac{e}{a+\delta} - \frac{E}{a+\delta}$$

$$= (E - e)\left(\frac{1}{a} - \frac{1}{a+\delta}\right) = \frac{(E-e)\delta}{a^2} \quad \ldots \ldots \quad (7).$$

The *capacity* has been altered from c to c' where

$$c' = \frac{E}{p'} = \frac{E\,a^2}{(E-e)\,\delta} \quad \ldots \ldots \quad (8).$$

    Let us now consider the *forces* acting :—

    At any point in the dielectric the charges $-E'$ and $+e$ have no effect, while the charges $+E$ and $-e$ can be considered to be collected at the centre of the ball; **therefore** the electric charge acts with a force $\dfrac{E-e}{a^2}$ on a unit of electricity concentrated at **any point of the** dielectric, for the charge E acts with a force $\dfrac{E}{a^2}$ and $-e$ with a force $\dfrac{-e}{a^2}$ neglecting quantities of the order $\dfrac{\delta}{a^3}$.

$\dfrac{E - e}{a^2}$ is the quantity which was denoted by R . . . . . (9)

in equation (1).

The dielectric moment of the whole dielectric is $e \delta$, for δ is its thickness and e the charge of one side. The volume of the whole dielectric is

$$4 \pi a^2 \delta.*$$

The dielectric moment of a unit of volume is therefore

$$\frac{e \delta}{4 \pi a^2 \delta} \quad . \quad . \quad . \quad . \quad (10).$$

This quantity was denoted by μ in equation (1).

The equation (1)

$$\mu = k\,\mathrm{R}$$

may therefore be now written by substituting from (10) and (9)

$$\frac{e}{4 \pi a^2} = k\frac{E - e}{a^2}$$

or

$$e = 4 \pi k\,(E - e) \quad . \quad . \quad . \quad . \quad (11),$$

whence we obtain

$$e = \frac{4 \pi k}{1 + 4 \pi k}\,E \quad . \quad . \quad . \quad . \quad (12),$$

and

$$E - e = \frac{E}{1 + 4 \pi k} \quad . \quad . \quad . \quad . \quad (13).$$

Substituting this value (13) in (8) we obtain for c'

$$c' = (1 + 4 \pi k)\frac{a^2}{\delta} \quad . \quad . \quad . \quad . \quad (14),$$

and substituting in (14) the value of $\dfrac{a^2}{\delta}$ from (5) we have

$$c' = (1 + 4 \pi k)\,c \quad . \quad . \quad . \quad . \quad (15).$$

But the ratio of the capacities equals K, the specific inductive capacity of the dielectric; and we have therefore

$$K = \frac{c'}{c} = (1 + 4 \pi k) \quad . \quad . \quad . \quad (16),$$

which was to be proved. We note that any small part of a condenser with flat plates may be considered as part of a spherical condenser whose radius is very large.

---

* For the volume of sphere of radius a, $= \dfrac{4}{3} \pi\, a^3$; that of one of radius $(a + \delta) = \dfrac{4}{3} \pi\, (a + \delta)^3$. Their difference, which is the volume of a shell of thickness δ, is $\dfrac{4}{3} \pi\, (3\, a^2\, \delta + 3\, a\, \delta^2 + \delta^3)$, and, as δ is small, we may neglect its second and third powers.

We are now to consider the attraction of a dielectric ball.

Let us place at any point, say for instance on the " working-ball," a charge of electricity + E, and let it attract a dielectric ball of radius *b*. Each volume element of the ball will be dielectrically polarized, and the side nearest to + E will be negatively charged, and that furthest from it will be positively charged.

In the following argument we must consider the distance between the charge (i.e. the " the working-ball") and the dielectric ball to be so great, compared with the radius of the latter, that all the lines of force may be taken as parallel, and that all points of the dielectric may be taken to be at practically the same distance from the charge, so that the forces acting on the electricity of the dielectric may be considered equal and parallel.

When the electricity of the dielectric ball is undisturbed, we may, if we please, consider every element of it to be charged with equal and opposite electricities. If the total amount of each throughout the ball be called *e*, we may regard the ball as composed of two coincident spheres uniformly charged throughout—the one, which we will call A, with negative electricity, the other, B, with positive; the total charge being in the one case — *e*, in the other + *e*. Since at present the two spheres are supposed to coincide, the result is a sphere at every point of which equal and opposite electricities are present: in other words, a perfectly neutral ball.

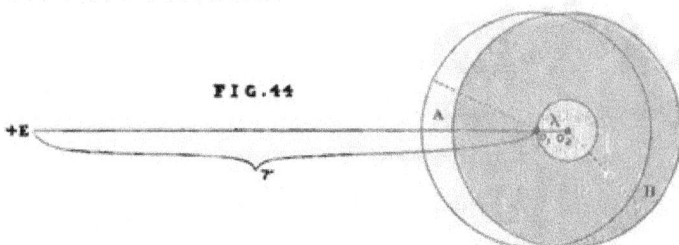

FIG.44

Now let us consider the effect of the polarization induced by the presence of the charge + E on the " working-ball." Since the forces exerted on the electricities at every point of the dielectric are to be regarded as equal and parallel, the equal and opposite electricities at every point of the dielectric will be separated to the same distance, and in a direction parallel to the line joining the charge + E with the centre of the dielectric. Regarding separately the resulting disturbance of all the positive and all the negative charges, we see that the total effect may be represented by supposing that the two spheres of positive and negative electricity which originally coincided and filled the dielectric ball are displaced as in fig. 44; the negative sphere, A, being moved slightly towards the charge + E, the positive, B, slightly away from it. The total force exerted by the now polarized dielectric may be regarded as the resultant of the forces exerted by these two spheres, A and B.

We can determine the action on the whole dielectric ball from the action on these two imaginary balls.

We have the one A containing a quantity of negative electricity — *e*. Let its centre be at O_1 (fig. 44). The second ball B contains a quantity $+ e$, and we will suppose its centre to be at O_2.

Let us put the distance O_1 $O_2 = \lambda$, where λ is a very small quantity.

The two balls A and B will now have the same external effect as if their charges $- e$ and $+ e$ were concentrated at their centres O and O_2 respectively.

The dielectric moment of the two balls A B is now

$$e \cdot O_1 \, O_2, \text{ or } e \, \lambda* \quad \dots \quad (17).$$

As the radius of each of these balls is b, the volume of each will be

$$\frac{4 \pi}{3} b^3 \quad \dots \quad (18),$$

and therefore the dielectric moment μ for each unit of volume will be got by dividing (17) by (18), and will be

$$\mu = \frac{3 e \lambda}{4 \pi b^3} \quad \dots \quad (19).$$

The electric attraction, on the charge E, of the ball A, with its centre at O_1, is

$$\frac{E e}{r^2} \quad \dots \quad (20),$$

where r equals distance E O_1.

The repulsion of the other ball B is

$$\frac{E e}{(r + \lambda)^2} \quad \dots \quad (21),$$

which gives for the total attraction h exercised by the charge E on the dielectric ball—

$$h = \frac{E e}{r^2} - \frac{E e}{(r + \lambda)^2}$$

$$= \frac{E e \left\{ r^2 + 2 \lambda r + \lambda^2 - r^2 \right\}}{r^2 (r + \lambda)^2}$$

$$= \frac{E e 2 \lambda r}{r^4} = E e \frac{2 \lambda}{r^3} \quad \dots \quad (22)$$

neglecting quantities of the order λ^2.

We shall now see the use of the equation (1), $\mu = k \, R$.

We know from (19) that

$$\mu = \frac{3 e \lambda}{4 \pi b^3}$$

and we wish to find R.

For this purpose let us imagine a unit of positive electricity collected at the point O_1, and see what is the total force exercised on it.

The charge $+ E$ exercises on it a repulsion—

$$R_1 = \frac{E}{r^2} \quad \dots \quad (23).$$

---

* This expression is the same as that for the moment of a magnet whose length is λ, and the magnetic strength of whose N. and S. poles is $- e$ and $+ e$ respectively. See page 149.

The ball A cannot act upon O_1 because O_1 is exactly at its centre.

Let us suppose the ball B to be divided in two parts.

One a *small* ball of radius λ, and having its centre at O_2 (the small circle in fig. 44). Let us call the electric charge of this, ϵ.

The other, a hollow ball, surrounding it.

This hollow ball does not act on O_1. The little ball, however, exercises a repulsion—

$$R_2 = \frac{\epsilon}{\lambda^2} \quad \ldots \quad \ldots \quad (24)$$

upon O_1.

As the ball B is supposed to be entirely and uniformly filled with electricity, we have, ratio—

$$\epsilon : e = \left\{ \begin{array}{c} \text{volume} \\ \text{of small ball} \end{array} \right\} : \left\{ \begin{array}{c} \text{volume of} \\ \text{ball B} \end{array} \right\},$$

or

$$\epsilon : e = \frac{4\pi}{3} \lambda^3 : \frac{4\pi}{3} b^3,$$

that is

$$\epsilon = \frac{e\lambda^3}{b^3} \quad \ldots \quad \ldots \quad (25),$$

whence, from (24)

$$R_2 = \frac{e\lambda}{b^3} \quad \ldots \quad \ldots \quad (26).$$

Since R_1 and R_2 act together, and in opposite directions, we have from (23) and (26)

$$R = R_1 - R_2 = \frac{E}{r^2} - \frac{e\lambda}{b^3} \quad \ldots \quad \ldots \quad (27).$$

This is the total force towards E which acts on the electricity concentrated at the point O_1.

The equation (1)

$$\mu = kR,$$

now becomes from (19) and (27)

$$\frac{3 e \lambda}{4 \pi b^3} = \frac{k E}{r^2} - \frac{k e \lambda}{b^3} \quad \ldots \quad \ldots \quad (28),$$

whence we obtain

$$e \lambda = \frac{4 \pi k b^3 E}{(4 \pi k + 3) r^2} \quad \ldots \quad \ldots \quad (29).$$

Let us now substitute in (29), the value of $1 + 4\pi k$ given in (16), and we shall have

$$e \lambda = \frac{(K - 1) b^3 E}{(K + 2) r^2} \quad \ldots \quad \ldots \quad (30).$$

Let us substitute this value of $e\lambda$ in the expression (22) which we found for h, the attraction of the dielectric ball, and we shall have

$$h = \frac{K - 1}{K + 2} \cdot \frac{2 E^2 b^3}{r^4} \quad \ldots \quad \ldots \quad (31).$$

To obtain the attraction h' of a metal ball, we put $K = \infty$ (see page 102) in the expression (31), which gives

$$h' = \frac{2\,E^*\,b^3}{r^3} \quad \ldots \quad (32).$$

The ratio (E^* page 92) of the attractions of metal and dielectric balls is then from (31) and (32),

$$E^* = \frac{h'}{h} = \frac{K + 2}{K - 1} \quad \ldots \quad (33),$$

and we see that (33) gives us

$$K = \frac{E^* + 2}{E^* - 1} \quad \ldots \quad (34),$$

the equation used on page 92.

In the actual experiments the correction,[†] which has to be applied in consequence of the balls not being indefinitely small, is exceedingly minute; the ratio of the corrected and uncorrected values of E^* is only that of 1 to 1·003 or 1·004, which would cause an absolutely inappreciable difference in the value of K.

---

[*] Of course, this is not the same E as we have been using in this Appendix.

[†] The corrections used will be found in Wiener Sitz., Bd. lxx., part ii., page 307.

DR. HOPKINSON'S EXPERIMENTS.

Since page 123 was written, Dr. Hopkinson has (on Nov. 3rd, 1880) communicated to the Royal Society [*] an account of some further experiments which he had made, in order if possible to account for the difference between his results and mine. He finds that, with a rough model of the 5-plate balance, varying the distance apart of the plates alters the results obtained. He suggests that some such change of position may account for the changes in the inductive capacity of glass observed by me. It is, however, certain that, having regard to the numerous minute precautions which have to be taken in working the 5-plate balance, no "rough model" could possibly give even approximate results; and further, although the balance certainly had been taken to pieces and put together again, between Christmas, 1877, and August, 1879, yet its construction is such that the distances apart of the plates cannot possibly have been appreciably altered.

[*] Proc. Roy. Soc., vol. xxxi. 1880-81, p. 148.

EXPERIMENTS ON LIQUIDS.

On January 6th, 1881, Dr. **Hopkinson communicated** to the Royal Society * the following determinations **of the** specific inductive capacities of certain liquids :—

Name of Liquid.	Specific Inductive Capacity.
Petroleum spirit (Field's)	1·92
Petroleum oil (Field's)	2·07
Petroleum oil (common)	2·10
Ozokerit lubricating oil (Field's)	2·13
Turpentine (commercial)	2·23
Castor oil	4·78
Sperm oil	3·02
Olive oil	3·16
Neats'-foot oil	3·07

* Proc. Roy. Soc., vol. xxxi. 1880-81, p. 347.

PART II.

MAGNETISM.

A PHYSICAL TREATISE

ELECTRICITY AND MAGNETISM.

---·---

Part II.

MAGNETISM.

CHAPTER XII.

MAGNETISM—PRELIMINARY EXPERIMENTS.

FOR these experiments there will be required two steel bar magnets about six inches long, a small compass, some knitting-pins and sewing-needles, a piece of watch-spring, some soft iron bars, and some iron filings, 2 or 3 lbs. As these can be procured for two or three shillings, the student is recommended to obtain them, and repeat the experiments as they are described.

Experiments.—It will be found that magnets attract pieces of iron or steel, and that the iron, and generally the steel, is attracted equally by either end of the magnet. It will also be noticed that, if we bring one of the magnets near the compass, one end of it will attract its north-pointing end and repel the opposite, while the other will attract the south-pointing end and repel that which points to the north.

Thus there is a difference in the ends of the magnets. To distinguish the ends it is the practice of European manfacturers to make a file mark on that end of the magnet which attracts the south-pointing end of the compass-needle. For brevity I will in future call this end the marked end of the magnet and the other the unmarked end.

L

Experiment.—Float one of the magnets on a bit of cork in a basin of water, so that it is free to move, and holding the other in the hand, bring first its marked and then its unmarked end near the marked and unmarked ends of the floating magnet.

It will then be found that

The marked end of one repels the marked end of the other and attracts the plain end, while the plain end repels the plain end and attracts the marked end; or, briefly, first defining that

The ends of a magnet are called its poles,

Like *magnetic poles* **repel** *each other,* **unlike attract.**

For the next experiment it will be necessary to manufacture some magnets.

MAGNETIZATION.

Magnetization may be performed by rubbing a bar of hard steel with a permanent **magnet in a** particular way, or rather any **one** of two or three ways. The simplest is called " magnetization **by single touch."** To perform this, the bar to be magnetized is laid **on a board,** fig. 45, near one end of which is a stop whose height **is less** than the thickness of the bar.

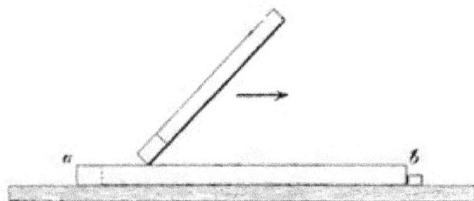

Fig. 45.

The magnet being held in a sloping position is drawn over the bar many times, always in the same direction, say from *a* to *b*, and always with the same end, say the marked one downwards. The bar will now become a magnet with its marked end at *a*. If either the direction of motion, or the pole used, had been changed, the magnetization of the bar would have been reversed. If both were changed, it would not be reversed, or rather would be reversed twice.

Another and more convenient method is called " magnetization by double touch," fig. 46. In this method the bar or knitting needle to be magnetized is laid on the table, and two magnets being taken, one in each hand, by opposite ends, the other ends, viz. the marked end of one magnet and the plain end of the other, are held together at the centre of the steel bar. They are now drawn apart quite over the ends, lifted, and

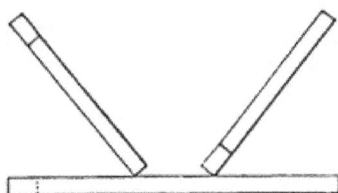

Fig. 46.

replaced at the centre, when the same process is repeated. After a number of repetitions, the steel bar is found to be a more or less powerful magnet, the marked end being that over which the plain end of the magnetizing magnet has passed, and *vice versâ.* Let us now take the magnet which we have made, and make a file mark on it, that we may know which is the marked pole.

Let a pen-tray, or other convenient vessel, be filled with iron filings, and the magnet put into it. On lifting it out, it will be found, if the process of magnetization has been skilfully performed, that a large bunch of iron filings hangs to each end, and that none, or hardly any, hang to any other part.

Thus the whole of the magnetic force of the magnet resides at its poles.

POLES INSEPARABLE.

Let us now try whether we can separate the poles, so as to have one magnet with only a marked pole, and another with only a plain pole.

By means of a file or pair of pliers let the magnetized bar be cut into two pieces ; it will be found that the old ends still retain all their magnetic properties unchanged, but the new end of the piece which had the file mark on it is an unmarked pole, and the new end of the other is a marked pole, and that either of the new ends will support as big a bunch of iron filings as either of the old ends.

STRENGTH OF POLES EQUAL.

Thus every magnet has two opposite poles. To show that the

L 2

strength of these poles is equal, magnetize a piece of watch-spring, and bend it round so as to bring the poles together (fig. 47). On bringing it near* to a compass, there will be no effect whatever, which shows that the attraction of the one pole exactly neutralizes the repulsion of the other.

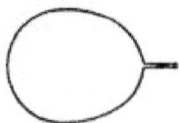

Fig. 47.

A better experiment, however, is the following. We know that the marked end of a compass needle points to the north and the plain end to the south; we can then consider the earth as a huge magnet, whose marked end is at the South Pole and whose plain end is at the North Pole.

Let a magnet be floated in water so that it is free to move in any direction. If the poles are not of equal strength, but say the marked pole is the strongest, it will be more attracted to the north than to the south, and will move northwards. Nothing of the kind actually occurs; the needle twists round till it has taken up a position pointing north and south, and then remains at rest. The opposite poles are of equal strength. In fact, if we take any magnet and cut it in two, each piece becomes a separate magnet, whose poles are of the same strength as the original poles, and generally :—

Into however many pieces we cut a magnet, each will have two opposite poles, whose strength is equal to that of the poles of the original magnet.

MAGNETIC FIELD.

Any region where forces act is called a field of force. If the forces are magnetic, it is called a field of magnetic force, or briefly a *magnetic field.*

When the direction and magnitude of the forces are equal at all points in the field, the field is called uniform; when the forces are magnetic, such a region is called a *uniform magnetic field.*

Throughout a region the size of any ordinary table, the earth's magnetic force may be considered uniform.

For convenience let the table be set so that the compass needle points along it. Now at all points just above this

---

* The poles must not be brought too close, or the one nearer the compass will act more powerfully than the one far from it.

table there will be a constant magnetic force parallel to the
length of the table. It will not be horizontal; but, as a constant
proportion of it will be horizontal, we may, for our present purposes, consider it to be so.

COUPLE.

The force on any compass-needle, or other magnet on the table,
is now what is called a *couple*.

Definition of a couple.—A couple consists of two equal and
opposite forces, acting on a body in directions which are parallel,
but do not coincide. It is obvious that a couple cannot move a
body, but only cause it to revolve. The force acting on a water-
wheel is a good instance of a couple. Neglecting for a moment the
weight of the wheel, we see that, if there were no axle, the force of
the water acting at A, fig. 48,
would simply carry the wheel
down stream. When, how-
ever, an axle is inserted at O,
then, in order to prevent
the wheel being carried down
stream, the bearings of the
axle have to press on it with
a force equal to that exerted

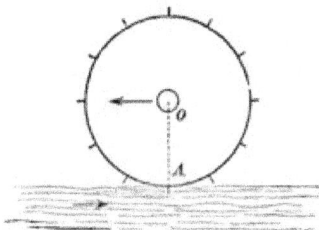

Fig. 48.

by the water, and in an opposite direction. Here there are
two equal forces forming a couple, and their effect is to make
the wheel revolve.

A magnet in a uniform magnetic field, in any position except
parallel to the lines of force, is subject to just such a couple,
only, if the magnet is free to move, it very soon takes up a
position parallel to the lines of force, and then the directions of
the two forces coincide and the couple disappears.

MOMENT.

The directive force of a magnet, in any position in any field,
equals the force which a torsion thread at its centre would have
to exert to keep it in that position, and is the *moment* of the couple
exerted by the magnet, or as it is generally called, the moment
of the magnet.

Definition of moment.—The moment of a couple is the strength
of either of its equal forces, multiplied by the perpendicular dis-

tance between the lines, which represent their directions and pass
through their points of application.

To return to our water-wheel illustration. The moment, or
turning **force** of the couple acting **on** the wheel, **is** the pressure
exerted **by** the water on the paddles multiplied by the radius (**or**
distance **from** axle **to paddle)** of the wheel. If we double **the**
water pressure, we double **the** turning force ; **or,** if we double **the**
radius **of** the wheel, the **same** water pressure will exert double
the turning force. If we halve the one and double the other, the
turning force will remain constant.

In districts where water is scarce, **the** necessary mill-power **is**
got by increasing the diameter **of** the wheels.

To FIND THE MAGNETIC MOMENT OF A MAGNET IN ANY POSITION.

Draw the length and position of the magnet and the direction
of **the lines** of force of the field, fig. 49.

Draw lines A E, **B D,** through the poles A B of the magnet

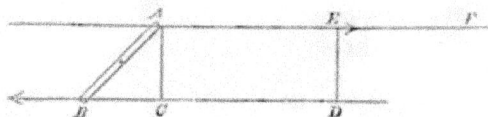

Fig. 49.

parallel **to the lines of force.** Draw A C perpendicular **to these**
lines. **Then A C, the distance between the** forces, is **called the**
arm of the couple.

The **length** A C, **multiplied by the strength of** the pole, is the
moment **of** the magnet A B in **its** present position.

The total moment of a magnet is its moment when its direction
is perpendicular to the lines of force ; **that is,** when **A C becomes**
equal **to A B.**

We know that the area of a rectangle **is its length multiplied
by its** breadth.

Let us now, on the **line A F, mark off a** length A E, **proportional to** the strength **of the pole, and** complete the rectangle
A C D E.

The **area** of this rectangle **is then equal** to the length A C
multiplied by the length A E, **or, in other** words, to the strength
of the pole multiplied by the arm of the couple; that is, the
rectangle A C D E is proportional to the moment of the couple.

Thus, when we have settled what length on the line A F shall represent a unit of magnetic force, we have a convenient method of finding the moment of any magnet in any position.

THE MOMENT OF A MAGNET IS NOT ALTERED BY CUTTING IT IN PIECES.

Let us take our magnet A B, and, having determined its moment in any given position, cut it up and determine the moments of the separate pieces when they lie in the same direction. On adding them we shall find they are together equal to the moment of the original magnet.

For, let the direction of the force be as before A F (fig. 50), and

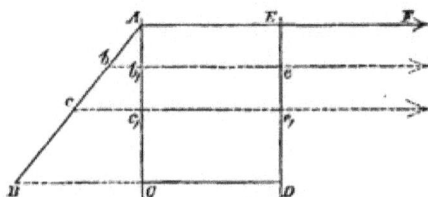

Fig. 50.

let the strength of the pole A be represented by the length A E, then, as before, A C is the arm of the couple and its moment is the area A C D E. Now cut the magnet into, say, three pieces, Ab, bc, cB, and determine their moments. To do this draw be, ce_1, in the direction of the force at b and c, viz. parallel to A E, and mark off on them lengths b_1 e, c_1 e_1, to represent the strength of the poles b c. By our experiment (page 148) we know that these are each equal to that of A, so that the lines A E, b_1 e, $c_1 e_1$, are all equal. The arms of the new couples are respectively A b_1, b_1 c_1, c_1 C, and therefore their moments are the areas

$$A \, b_1, \, e \, E, \quad b_1 \, c_1, \, e_1 \, e, \quad c_1 \, C \, D \, e_1,$$

which added together make up the area A C D E, the moment of the original magnet.

The total moment of a magnet is the moment when it is at right angles to the lines of force.

MAGNETIC POTENTIAL.

The potential due to a given magnetic pole is of precisely the same nature as that due to an electrified body at that place. The

whole of the chapter about electric potential (page 26) may be
applied to magnetic poles, with the following alteration:—

For $\left\{\begin{array}{c}\text{Body charged with}\\ +\text{ electricity}\end{array}\right\}$ read $\left\{\begin{array}{c}\text{marked magnetic}\\ \text{pole.}\end{array}\right\}$

For $\left\{\begin{array}{c}\text{Body charged with}\\ -\text{ electricity}\end{array}\right\}$ read $\left\{\begin{array}{c}\text{unmarked magnetic}\\ \text{pole.}\end{array}\right\}$

If, however, we wish to obtain the whole system of equipotential surfaces due to a magnet, we must remember that they
will be " symmetrical" on the two sides of a line drawn at right
angles to and at the middle point of the straight line joining the
poles. For, suppose a magnet A B to be of any shape, fig. 51, the

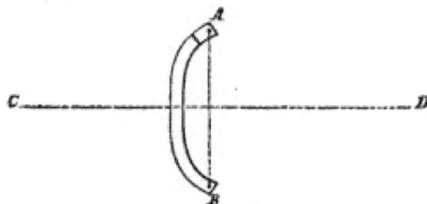

Fig. 51.

line of symmetry will be the line C D in the figure. When we say
that the system will be symmetrical with regard to the line C D, we
mean that, if we draw the system for one side, say the side A, and
then place a looking-glass along the line C D, the reflection of
the drawn system will be a correct representation of the system
on the side B.

Corollary.—The potential along the line C D is zero (see pages
27, 28), for the potential at any point of it is the potential due
to two equi-distant equal poles of opposite signs.

MAGNETIC INDUCTION.

Experiment.— Take a short bar of soft iron *a b*, fig. 52, of the

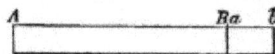

Fig. 52.

same shape as one of the magnets and place it against one pole;
it is attracted, and either end of it is attracted equally.

Now bring the compound bar A *b* near to the compass. We
shall find that there is hardly any attraction or repulsion at B,
but that a new pole almost equal in strength but of opposite kind
to A has been developed at *b*. This shows us that the end B of

the steel magnet has induced at *a* a pole of opposite nature and almost equal strength to itself.

This accounts for the attraction of the soft iron, as the pole B attracts the opposite pole which it has itself induced. (Compare page 7.)

INTENSITY OF MAGNETIZATION.

The same steel bar may be magnetized more or less intensely.

As a rule, if we increase the strength of the magnetizing magnet, we increase the magnetization of the bar. In all bars there is, however, a certain limit, after which no amount of magnetic force can increase their permanent magnetism. This is called their "saturation point," and bars so magnetized are said to be magnetized to saturation.

It is possible to super-saturate a bar with magnetism, that is, to temporarily give it a stronger magnetization than it can permanently retain. It is then found that, after the inducing magnetic force is removed, the magnetic force diminishes at a gradually decreasing rate until it has reached its permanent amount. That is, for the first few hours it diminishes rapidly, then more slowly for some days, and very slowly for many weeks. For this reason all magnets used in investigations, where the intensity of magnetization is assumed constant throughout the experiments, should be magnetized at least six months beforehand.

The removal of the super-saturating magnetism may be hastened by any process which tends to allow the molecules to slide over each other. Such, for instance, as alternate cooling and heating in cold water, and water, say, at 150° F., gentle blows with a hammer, &c.

M. Jamin [*] has constructed two magnets, one whose weight is 6 kilogrammes (about 12 lbs.), and which will support 80 kilogrammes, or 13⅓ times its own weight; another weighing 50 kilogrammes, which will support 500. This last is by far the most powerful permanent magnet which has yet been constructed.

The above weights were supported after the magnets had attained their permanent condition.

Another of M. Jamin's magnets carried sixteen times its own weight immediately after being magnetized.

---

[*] Comptes Rendus, 1873, T. lxxvi. p. 1153.

It is said that very small magnets have been constructed which will carry twenty-five times their own weight.

MAGNETIC UNITS.—UNIT MAGNETIC POLE.

A unit magnetic pole is **a pole** which repels another unit pole at unit distance with unit of force.

In the C.G.S. system* it is the pole which repels a similar pole distant 1 centimetre with a force of 1 dyne.

UNITS.—MAGNETIC MOMENT.

The unit of magnetic moment is the moment of a magnet of unit length, the strength of whose poles is equal to unity; or generally of any magnet, the product of whose strength into its length is equal to unity.

In the C.G.S. system the magnet whose moment is unity is a magnet whose length is 1 centimetre, and the strength of whose poles has the unit value defined in the last paragraph.

The magnetic moment of a magnet is the strength of one of its magnetic poles multiplied by the length of its axis. That of the earth is equal to

$$85,500,000,000,000,000,000,000,000 \text{ C.G.S. units.}$$

INDEX NOTATION.

This is a good opportunity to give the notation now adopted to save writing a long list of noughts in large numbers. It consists in writing the first of the whole numbers followed by a decimal point, and then indicating that the whole number and decimal are to be multiplied by 10 as many times as may be necessary. The number of times is indicated by a small figure called the "index" above and after a 10.†

This index gives the number of noughts in the multiplier; and as only one figure is put to the left of the decimal point in the number, the index is also the total number of noughts and figures after the first number.

Thus, 8.5×10^6 is known at once to be 8 followed by 6 figures, the first of which is a 5 and the rest noughts.

---

* Page 40.

† Thus, $10^2 = 10 \times 10 = 100$,
$10^3 = 10 \times 10 \times 10 = 1000$, and so on.

In this system of notation the earth's magnetic moment is written

$$8{\cdot}55 \times 10^{25}$$

which is a much more convenient number both to write and read than the row of noughts above. It shows that the decimal point is to be moved 25 figures to *the right*. Long decimals are expressed in a similar way, only the index is written with a negative sign. A negative index means that the number is to be *divided* by 10 that number of times.

Thus,

$$8{\cdot}55 \times 10^{-6}$$

means a decimal which, if written in full, is

$$0{\cdot}00000855$$

It shows that the decimal point is to be moved 6 figures to the left.*

UNITS.—INTENSITY OF MAGNETIZATION.

Let us imagine a quantity of a substance all uniformly magnetized, and let us cut any bar out of it and determine its magnetic moment.

Let us now, leaving the section of the bar the same, double its length. We shall double its magnetic moment, because, leaving the strength of the poles the same, we have doubled the arm of the couple. In doubling its length we have doubled its volume.

Again, leaving the length the same, let us double the section; we have now also doubled the magnetic moment, for the strength of the poles has been doubled and the arm of the couple has remained constant. In this case also we have doubled the volume.

So again, if we double both the length and the cross section, we quadruple both the moment and the volume.

Hence we may state,—

If from any uniformly magnetized substance we cut any piece whatever, its magnetic moment is simply proportional to its volume.†

We can now define the unit intensity of magnetization.

The intensity of magnetization of any uniformly magnetized

---

* It may be noted that this index is also the index or characteristic of the logarithm of the number.

† We see that, in fig. 50, p. 151, if we cut the magnet in two, and lay the pieces side by side, we shall have the area of the rectangle A C D E still the same, because we shall have halved its height and doubled its length.

substance is unity when a unit of volume of the substance has a
magnetic moment equal to unity.

In the C.G.S. system the unit of intensity of magnetization is the
intensity to which a substance must be magnetized that a cubic cen-
timetre of volume of it may have a magnetic moment equal to unity
as defined above. (Page 154.)

The intensity of magnetic field is the force which a unit pole
will experience when placed within it.

A magnetic field of unit intensity is in the C.G.S. system a field
where a unit pole will be acted on with a force of one dyne.

MAXIMUM OF PERMANENT MAGNETISM.

It is stated by Kohlrausch[*] that the maximum of magnetization which long thin steel rods can permanently retain is about

785 C.G.S. units of intensity.

PULL ON A COMPASS NEEDLE.

The horizontal force of the earth's magnetism at Kew is now
about

·179 dyne.

That is, if a compass needle, the strength of whose pole is unity,
be placed E. and W., the force acting on each pole, and tending
to turn it to an N.S. position, will be ·179.

If the needle were one of Kohlrausch's magnets, the force
would be

·179 × 785 = 90·4 dynes.

The force of gravity on a gramme mass at Greenwich is about
981, and therefore, if the needle were kept in the E.W. position
by means of a cord passing over a pulley, then the weight at the
bottom of the cord would have to be $\frac{90\cdot4}{981}$ gramme for each pole
in order that the earth's force might be exactly balanced; that
is about $\frac{9}{10}$ gramme for each pole and 1.8 gramme for the whole
magnet. 1·8 gramme is about 27 grains.

MAGNETIC SOLENOIDS AND SHELLS

For the purposes of mathematical calculation, actual magnets
are often supposed to be replaced by certain imaginary magnets

[*] " Physical Measurements," English Edition, p. 195.

equal to the actual magnets with respect to the points under observation, but, owing to their shape, easier to use for purposes of calculation.

The imaginary magnets are supposed to be made of what is called " Imaginary magnetic matter;" that is, of a substance which has magnetic properties and no others, such as weight, volume, &c.

For instance, we can consider that any quantity of this matter can be collected *at* the same mathematical point.

DEFINITIONS.*

(1.) " *A magnetic solenoid* is an infinitely thin bar of any form longitudinally magnetized with an intensity varying inversely as the area of the normal section [that is, the cross section perpendicular to the length] in different parts."

[When of two quantities one varies inversely as the other, the product is constant.]

" The constant product of the intensity of magnetization into the area of the normal section is called the magnetic strength, or sometimes simply the strength of the solenoid. Hence the magnetic moment of any straight portion or of an infinitely small portion of a curved solenoid is equal to the product of the magnetic strength into the length of the portion."

(2.) " A number of magnetic solenoids of different lengths may be put together so as to constitute what is, so far as regards magnetic action, equivalent to a single infinitely thin bar of any form longitudinally magnetized with an intensity varying arbitrarily from one end of the bar to the other. Hence such a magnet may be called a complex magnetic solenoid.

" The magnetic strength of a complex solenoid is not uniform, but varies from one part to another."

(3.) " An infinitely thin closed ring magnetized in the manner described in (1) is called a closed magnetic solenoid."— (Thomson.)

Now it can be shown mathematically that the potential due to

---

* See Professor **J.** Clerk Maxwell, *Electricity*, **vol. ii.** chap. iii., and Sir W. Thomson, *Mathematical Theory of Magnetism, Papers on Electrostatics and Magnetism*, **chap. v. p. 378.**

a simple solenoid, and *consequently all its magnetic effects*, depend only on its strength and the positions of its ends, and not at all on its form, whether straight or curved, between the ends.

That is, that a solenoid* can be considered as consisting of two equal portions of magnetic matter collected at the points where its ends are, and connected by a perfectly rigid non-magnetic bar without weight.

Now if a solenoid forms a closed curve, the potential due to it is zero at every point, for we have shown that there is no action due to any part except the poles, and here the poles exactly neutralize each other.

MAGNETIC SHELLS.—DEFINITION.

"If a thin shell of magnetic matter is magnetized in a direction everywhere normal [perpendicular] to its surface, the intensity of the magnetization at any place multiplied by the thickness of the shell at that place is called the strength of the magnetic shell at that place. If the strength of the shell is everywhere equal, it is called a simple magnetic shell; if it varies from point to point, it may be considered to be made up of a number † of simple shells [of different areas] superposed and overlapping each other. It is therefore called a Complex magnetic shell." (Maxwell.)

It can be proved mathematically that

The potential due to a magnetic shell at any point is the product of its strength into the solid angle subtended by its edge at the given point.

SOLID ANGLE.

The plane angle subtended by a length at a point is, as we know, the angle between two lines drawn from the point to the extremities of the length, and this may be measured by describing a circle of unit radius round the point (fig. 53), and then the part of the circumference cut off between the lines is the angle required.

Fig. 53.

---

* When a "solenoid" is spoken of, a "simple solenoid" is meant, unless the contrary is stated.

† When a magnetic shell is spoken of, a simple magnetic shell is meant, unless the contrary is stated.

We see that if either the length increases or the point approaches nearer, the angle thus measured increases, and that it decreases if the length diminishes or the distance increases.

The definition of solid angle is analogous. The solid angle subtended at a point by the area bounded by any closed curve is thus measured:—Draw straight lines from all parts of the boundary curve to the point. These lines will then form a cone (fig. 54). Now, with the point for centre, describe a sphere of unit radius. The straight lines passing through its surface will mark out a curve on it more or less similar to the given boundary curve. *The area of this curve is the solid angle subtended at the point by the given area.*

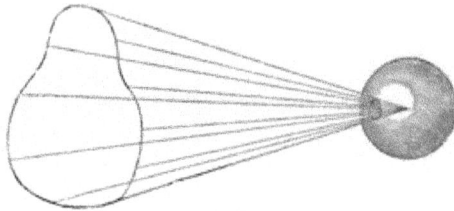

Fig. 54.

We see, if we increase the area, the lines will widen out, and therefore the solid angle increases, or, if we bring the point nearer, the solid angle increases.

As an illustration, let us place a candle at the point and a large screen at a fixed distance from it. For trying the experiment accurately, the screen would have to be a hollow sphere, but a flat one is quite sufficient for purposes of illustration. If nothing interposes, the whole of the light falls on the screen. If a piece of card is interposed, the light is divided between the card and the screen, and the size of the shadow of the card measures the quantity of light received by the card. The quantity of light received by the card—that is, the size of the shadow on the screen— is proportional to the solid angle subtended by the card at the point where the light is. It will be seen that a larger card increases the size of the shadow, or that moving the same card nearer the light has the same effect.

Again, if the card be turned more or less edgeways to the light, the shadow will be diminished or increased as the solid angle diminishes and increases.

EXPERIMENTAL DETERMINATION OF THE LINES OF MAGNETIC FORCE.

The directions of the forces emanating from a magnet are, like the electric forces, always perpendicular to the equipotential surfaces. The following is a method of tracing these lines of force directly.

Take a board of the same or greater thickness than the magnet, and let the magnet into it so that its upper side is flush with the upper surface of the board.

Lay a piece of smooth paper over all and fix it.

Now with a sieve or "colander" dust fine iron filings all over the paper.

As each falls near the magnet, it is magnetized by induction and turns round so as to lay its longest diameter in the line of force at that point.

Each of these minute magnets attracts the next till a continuous chain is formed all along each of the lines of force. The board must be tapped from time to time to overcome the friction of the filings on the paper.

If it be desired to preserve the curves, a piece of card with its under side gummed may be laid upon them. When the gum is dry, the filings stick to the card. The magnet may be placed under a piece of glass instead of paper if preferred, in which case it will not be necessary to let it into a board.

A good way to prepare these curves for exhibition in a magic lantern is to coat a piece of glass with some transparent cement,* which melts on being heated. When it is quite hard, a magnet must be placed under the glass, and the filings dusted on. The glass must then be carefully carried to an oven, warmed till the cement is soft, when the filings will sink in. On removing the glass and allowing it to get cold, the filings will be all fixed in position.

Plate XI. is an engraving of the actual lines of force determined by Faraday for different magnets.†

---

* The cement sold under the name of "Mend all" answers very well. It may be diluted with water for the purpose.

† See Chapters XXXVII., XXXVIII., and plates in them.

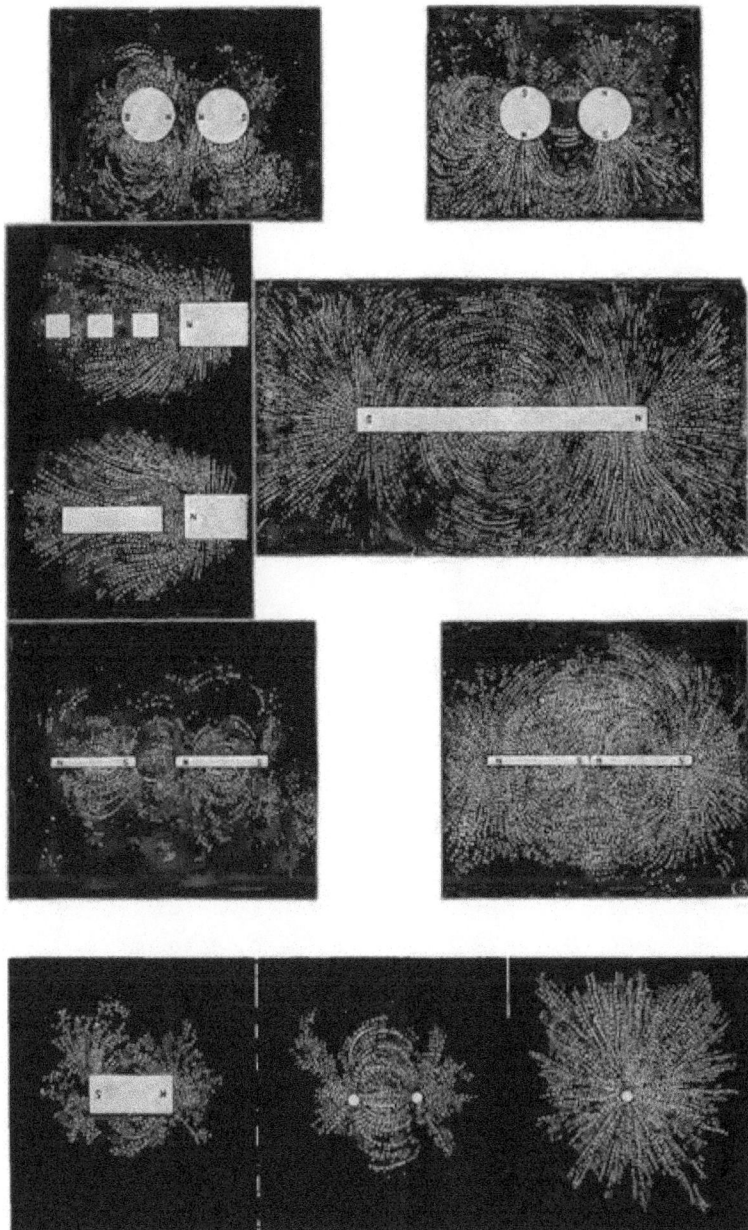

PLATE XI.—LINES OF MAGNETIC FORCE.

CHAPTER XIII.

INTRODUCTORY.

IF a magnet be balanced on a vertical pivot so that it is free to move in a horizontal plane, like a compass needle, it will take up a determinate position. The marked end points towards the north. Again, if a steel bar be exactly balanced on a horizontal axis and then magnetized, it will be found that it no longer balances in a horizontal position, but that (in the northern hemisphere) the marked end *dips*, the dip being least when the needle is in that position which it would have taken up if free to move in a horizontal plane.

The whole direction of the magnetic force may be seen at once by suspending a knitting needle by a fine thread attached to its middle, so that when unmagnetized it balances in a horizontal position. If now magnetized it will turn till the marked end points northwards and downwards (not vertically but at a certain angle) and there rest.

In different localities, or at different times in the same locality, such a needle does not always lie in the same direction.

It does not in general point due north.

The angle which the marked end * makes with a horizontal plane is called the "Inclination" or "Dip." The angle which that part of a vertical plane through the needle which contains the marked end makes with a line drawn due north from the centre of the needle is called the "Declination." Nautical men also call this angle the "Variation." We shall, however, use the word Declination for the above-mentioned angle, and confine the word Variation to changes in that angle.

This angle is of course the same as that which the direction of

---

* The marked end is specified to avoid ambiguity about + or — angles.

M

an ordinary horizontal compass needle makes with the direction of true north.

Thus to obtain the true bearings of an object from the compass it is necessary to allow for the declination * of the place at the time of observation. This will be found by referring to the Nautical Almanac. The present declination at Greenwich is about 18° W., that is, the marked end of the needle points 18° West of North.

EARLY OBSERVATIONS—DECLINATION.

The declination is believed to have been first noticed in Europe by one Peter Adsiger † in 1269; he found it to be then 5° E. The authenticity of the letter on which his claims rest is, however, very doubtful, and Humboldt considered that the first discovery of it was made by Columbus on September 13th, 1482.

Whether Columbus was or was not the first to observe the existence of the declination, there is no doubt that he was the first European who observed that the declination itself varied, and was different at different times and places. It was, however, known to the Chinese in the beginning of the twelfth century.

The first known work for the use of seamen was by Boroughs, Comptroller of the Navy to Queen Elizabeth. It was published in 1581 and reprinted in 1585. It is called

"*A discourse on the variation of the Campas or Magneticall Needle,*" and is dedicated to "*the travaillers, sea-men and mariners of England.*"

Boroughs found that the declination of the needle was, at Limehouse, on October 16th, 1580, equal to 11° 18′ E.

In the year 1622 Professors Gellibrand and Gunter determined to repeat Boroughs's observations at Limehouse. They found that the declination was then only 6° 15′, and Boroughs's reputation suffered, as he was believed to have made a mistake of nearly 50 per cent.

In 1634, the two Professors having obtained new and improved

---

* In using a pocket mariner's compass, before allowing for declination, it is well to look at the under side of the card, as such instruments are often corrected for Greenwich by fixing the magnet at an angle of 18° to the N.S. line on the card, i.e., N.N.W. and S.S.E. nearly.

† *Terrestrial and Cosmical Magnetism,* by E. Walker, M.A. (Adams Prize Essay, 1865). The whole of the early history of the declination in this chapter is taken from this work.

instruments, determined to give Boroughs's reputation another chance, and to repeat their observations. This they did, and found that the declination was now only 4° 4' 49" E.

Professor Gellibrand concludes his essay as follows: [*]—

"It were needlesse having so many sufficient testimonies to produce any more. . . . Hence we may conclude that for the space of 54 years (the difference of time between Mr Boroughs's and these last observations of ours) there hath beene a sensible diminution of 7° and better."

Mr. Walker goes on to say—

" From this period then the fact of the secular variation of the declination may be considered to be completely established, and the determination of the extent and the law of this variation became a recognized part of all magnetic investigation.

" From recorded observations we find that this diminution in the easterly declination which was called into notice by Gellibrand still continued, and that in 1660 or thereabouts the declination in London vanished, i. e. the magnetic needle pointed due N. and S. After this it [the marked end] began to deviate to the W., and this westerly declination went on increasing till it reached its maximum about the year 1818, when it attained the value of 24° 30' W. Since that time it has been decreasing, that is, the secular variation is now easterly."

THE INCLINATION.

THE DIP OR INCLINATION was discovered accidentally in 1576, by one Norman, an instrument maker. He published an account of it in a work which he called " The New Attractive," and from which Mr. Walker has extracted the following paragraph: [†]—

" Having made many and diverse compasses, and using alwaies to finish and end them before I touched the needle, I found continuallie that, after I had touched the yrons with the stone, that presentlie the North point thereof would bend or decline downwards under the horizon in some quantitie; insomuch that to the flie of the compass, which was before levell, I was still constrained to put some small piece of ware on the south point and make it equall againe. Which effect having many times passed my handes without anie great regard thereunto, as ignorant of

---

[*] Walker, Ibid., p. 15.
[†] Walker, Ibid., p. 146.

anie such propertie in the stone, and not before having heard **or**
read of anie such matter, it chaunced at length there came to my
hands an instrument to be made with a needle of sixe inches long,
which needle after I had pollished, cut of a just length and made
it stand levell upon the pin, so that nothing rested but onlie the
touching it with the stone, when I had touched the same, presentlie the North part thereof declined down in such sort that
being constrained to cut **awaie some** of that part to make it equall
againe, in the end I **cut it too** short, and so spoiled the needle
wherein I had taken so much pains.   Hereby being stroken into
some cholar I applied myself to seeke further into this effect, and
making certaine learned and expert men, my friends, acquainted in
this matter, they advised me **to** frame some instrument to make
some exact triall how much the needle touched **with the** stone
would decline, or what **greater angle** it would make with the
plaine of the horizon.

" Whereupon I made diligent proofes, the manner whereof is
showed in **the** chapter following :—

" **Take** a small needle of steele wier of five or six inches long,
the smaller and the finer mettall the better, and in the middle
thereof (crosse the same) by the best means you can, fixe as it were
a small axel tree of yron **or brasse of** an inch long or thereabout,
and make the ends thereof verie sharpe, whereupon the needle may
hang levell and plaie at his pleasure.   Then provide a round plaine
instrument like an astrolobe, to be **divided** exactlie into 160 (?)
parts whose diameter **must be the length of** the needle, or **thereabout**, and the same instrument to bee placed uppon a foote of
convenient height, **with a** plumme line to set it perpendicular.
Then in the centre of the same instrument place a piece of glasse
hollowed, and against the same centre uppon some plate of brasse
that may be fixed upon the foote of the instrument, **fit** another
piece of glasse in such sort that the sharpe endes of the axel tree
being borne in these two glasses, the needle may plaie freelie
at his pleasure according to the standing of the instrument; and
the **needle must be so** perfected that it may hang upon **his**
axel **tree, both** endes levell with the horizon, or being turned
may stand **and remaine** at anie place that it shall be set;
which being **done touch** the **sayd** needle with the magnet
stone, **and set the** instrument **perpendicular** by the plumme
line, **and turne the edge South and North so as** the needle

may stand duelie according to the variation of the place: which variation the needle of his own propertie would shewe, were it not that he is constrained to the contrarie by the axel tree.

"Then shall you see the declination of the North point of the touched needle, which for this citie of London I find by exact observations to be 71° 50′."

It was soon found that the dip is not the same in all places or at all times; that it is subject to variations in the same way as the declination is.

THE MAGNETIC EQUATOR.

Generally speaking, the needle is nearly horizontal at the equator. The points where it is exactly horizontal form an irregular curved line near the equator, and at the north of this line the marked end dips, at the south the plain end. This line is called the *magnetic equator.* As we approach either pole the needle becomes more and more nearly vertical, and at certain points in each hemisphere it stands exactly vertical.

THE INTENSITY.

The intensity of the earth's magnetic force is also different at different times and places.

The first attempt at systematic observation of it was made by the French Government in 1785. The expedition they sent out consisted of two frigates, *L'Astrolobe,* Capt. De Langle, and *La Boussole,* Capt. De La Perouse, commander-in-chief of the expedition. Letters dated Botany Bay, 1788, were received, and after that date nothing more was ever heard of the expedition. It is supposed that the ships were lost with all on board.

M. Paul de Lamanon, who accompanied the expedition, discovered that the force was less in the tropics than towards the poles; but the first systematic observations were those of Humboldt, made during his travels in America between 1798 and 1803.[*]

* Walker, Ibid., p. 183.

CHAPTER XIV.

TERRESTRIAL MAGNETISM.—MODERN EXPERIMENTAL METHODS.

WE shall now describe the methods used in modern observatories for continuous observation and registration of the three quantities—

Declination,
Inclination,
Intensity.

There are in practice four quantities to be determined, as the intensity of the force is always resolved in two directions at right angles to each other, and observations are made on

Horizontal Intensity, and
Vertical Intensity.

Of the three quantities, inclination, horizontal force, and vertical force, it is, however, only necessary to observe two, as when any two are known the third can be calculated. In survey work the inclination and the horizontal force are observed and the vertical force calculated. In observations where self-recording instruments are used, the vertical and horizontal force are observed and the inclination calculated.

INSTRUMENTS.

THE UNIFILAR MAGNETOMETER, KEW PATTERN.
Plates XII. and XIII.

This instrument is used
For determining the absolute horizontal force—
By observations of deflection.
By observations of vibration. And
For determining the declination, that is, the angle between the astronomical meridian and the direction of the earth's magnetic force.

DETERMINATION OF HORIZONTAL FORCE.

Plate XII. shows the instrument arranged for deflection observations.

Plate XIII. shows it arranged for observations of vibration.

We will immediately describe the experimental methods of determining the time of vibration accurately, and of making deflection observations; but we will first show how the value of the horizontal force can be determined by a combination of the observations of deflection and vibration.

Vibration Experiments.—It can be proved that, if a suspended or pivoted magnet be allowed to swing in a horizontal plane under the influence of the earth's force, the *product* of the earth's horizontal force into the magnetic moment of the magnet will be inversely as the square of the time of a vibration. The magnetic moment of a magnet is, as we know, its length multiplied by the strength of one of its poles.

Deflection Experiments.—If a suspended magnet be deflected by another magnet, the amount of deflection will depend on the *ratio* of the earth's horizontal force to the moment of the deflecting magnet.

In the observations which we are about to describe, the same magnet is vibrated, and is used as a deflector.

Let H be the horizontal force, m the magnetic moment of the deflecting and vibrating magnet K. These are unknown quantities to be determined. From the observations of deflection we can calculate the ratio of H to m; that is, if we put

$$\frac{H}{m} = A$$

A is a known quantity.

From the vibration observations we know the product of H and m; that is, if we put

$$H m = B,$$

B is a known quantity.

But if A and B are both known their product A B is also known.

Hence we have

$$H m \cdot \frac{H}{m} = A B$$

a known quantity.

Cancelling, we have

$$H^2 = A B.$$

or

$$H = \sqrt{A B},$$

that is, the horizontal component of the earth's magnetism equals the square root of the product of the two quantities determined respectively by observations of deflection and vibration.

Similarly, if we wish to determine m, we have

$$\frac{H m}{\frac{H}{m}} = \frac{B}{A} = m^2,$$

or

$$m = \sqrt{\frac{B}{A}}.$$

When we know m we can determine H at different times and places by observations of deflection only; so that it is only necessary to make the vibration observations at the beginning of a survey. It is better, however, to repeat them whenever there is a convenient opportunity, so as to guard against the introduction of error by accidental changes in the strength of the magnet.

OBSERVATIONS OF DEFLECTION.

Plate XII. shows the instrument arranged for deflection observations.

The instrument, when in use, is supported on a tripod stand similar to that used by photographers, three radial V grooves carrying the three levelling screws which form the feet of the instrument. The magnet G is suspended by some fibres of unspun silk C to a torsion head, and rack apparatus F, by which latter it can be raised and lowered.

Before suspending the magnet a circular brass plummet (E, Plate XIII.) of the same weight as the magnet is suspended for a considerable time to the thread and spins round under the influence of the torsion till that is eliminated. The plummet is then removed from its socket and the magnet attached instead, the socket being meanwhile held so that it may not twist and introduce fresh torsion.

A graduated brass bar, D, carries on a carriage, L, the deflecting magnet, K. Before K is put into position the azimuth of the suspended magnet is observed. K then causes a deflection. The

PLATE XII.—KEW UNIFILAR MAGNETOMETER, ARRANGED FOR DEFLECTION.

strength of K is constant and known, and it acts against the earth's horizontal force, which is variable. The amount of deflection is then a measure of the earth's horizontal force; when the force is great the deflection is small, when it is small the deflection is great.

The amount of deflection is observed by a modification of the mirror method already described.*

This modification consists in substituting for the concave mirror a plane mirror, H, and for the lamp a telescope, A, to which the scale B is fixed. On looking into the telescope part of the scale B is seen by reflection in the mirror. If the magnet is deflected the mirror turns with it and a different portion of the scale appears in the field of view. The eye-piece of the telescope is furnished with a vertical spider line, which enables the exact deflection to be measured; for the difference of two readings of the divisions of the scale crossed by the spider line gives the number of scale divisions corresponding to the deflection of the needle which has taken place between the readings.

Details as to the use of the Instrument.—It is first carefully levelled, and then, the deflecting magnet K being removed, the whole apparatus is turned on a vertical pivot (the end of which can be seen under the base) until the reflection of the centre of the scale B comes upon the spider lines in A. This is to ensure that the bar D shall be at right angles to the direction of the magnet. The apparatus is then clamped by means of the vertical screw M, and, if necessary, a fine adjustment is made by the horizontal tangent screw below M. The magnet K is then put on its carriage L, and slid along the bar to a given distance from the suspended magnet.

The deflection is then read.

The magnet K is then placed at an equal distance on the other side of the suspended magnet, and the deflection in the opposite direction observed. This eliminates any inaccuracy in the first adjustment of the instrument, as the second observation would, by such inaccuracy, be made as much too small as the first is too great, and *vice versâ*. When time permits, several pairs of observations at different distances are made.

Observations are also made with the deflecting magnet K in the same position but with its direction reversed.

* See Part I., p. 38.

From these observations we get, by a mathematical process, the ratio of the magnetic strength of the magnet K to the horizontal force of the earth.

In all the observations wooden shutters are inserted to protect the suspended magnet from currents of air.

The strength of the deflecting magnet K decreases as the temperature rises. The temperature must therefore be observed, the correction calculated, and the results of the observations all reduced to a uniform temperature.

The strength of the suspended magnet does not enter into the calculation, for any alteration in it would affect equally the earth's attraction and that of the deflecting magnet.*

OBSERVATIONS OF VIBRATION.†

The arrangement of Plate XII. *may* be used, the deflecting magnet being suspended in place of the former suspended magnet. A conspicuous point being marked at the zero of the scale, the time of a half-vibration is the interval between two successive passages of this point, as seen by reflection in the mirror across the spider line of the telescope.

This method is often used in determining times of vibration with other kinds of apparatus. With the Kew magnetometer it is however preferred to make the vibration observations' with the modification of the above arrangement, which is shown in Plate XIII.

For this purpose the telescope and scale, deflecting bar, and torsion apparatus are removed, and another telescope B, torsion apparatus D P H, and magnet box A are substituted, as shown in Plate XIII.

The magnet which is suspended is the same magnet as is marked K in Plate XII., and there used as a deflector. It consists

Fig. 55.

of a magnetized steel tube, fig. 55, at one end of which is flat glass, on which a minute scale is photographed, and at the other

* Part III., Chapter XX.

† See Maxwell's *Electricity*, Art. 456, vol. ii. p. 102, and *Admiralty Manual Instructions for Magnetic Surveys*, p. 19.

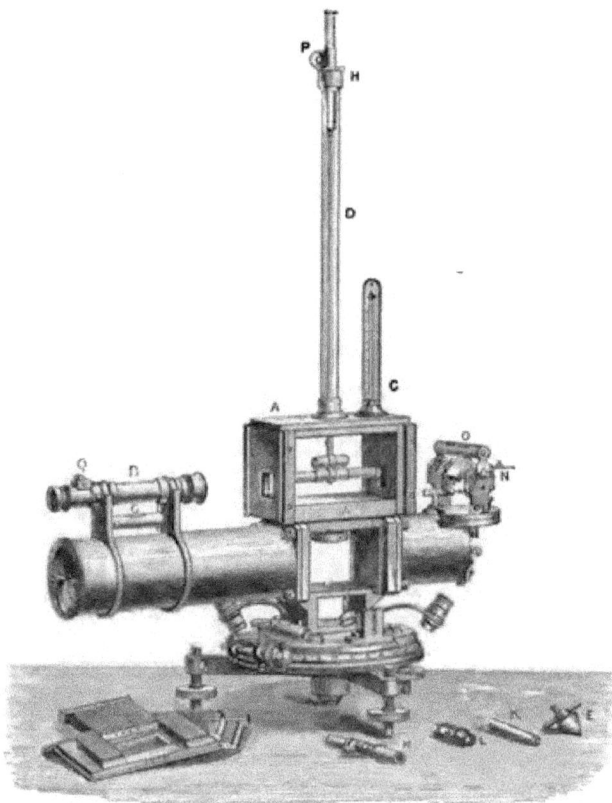

PLATE XIII.—NEW UNIFILAR MAGNETOMETER, ARRANGED FOR VIBRATION

is a collimating lens, that is, a lens such that rays diverging from the scale at one end of the magnet, and falling on the lens at the other are rendered parallel. The circle is now turned till the vertical spider line of the telescope B cuts the middle division of the scale of the magnet as the telescope looks into and through the magnet.

The magnet being set swinging, the approximate time of a half-vibration is roughly determined by simple observation. By the time of a whole vibration is meant the interval between two successive passages of the zero point *in the same direction* over the spider line. By the time of a half-vibration is meant the interval between two consecutive passages; the consecutive passages being, of course, in opposite directions. In some French books the latter, which in England is called a half-vibration, is spoken of as a vibration.

The time of a half-vibration will be, with the magnets usually supplied, from 2 to 4 seconds.

A clock beating seconds being placed where it can be seen and where the beats can be heard, the time at which the observer removes his eye from the clock is called out and written down by an assistant; the observer goes on counting by ear from the time when he looks away from the clock, and looking through the telescope notes between which beats the centre of the scale crosses the spider line. The fractions of a second are determined by estimation and observation of the scale. If, for instance, at the beat before the passage the spider line was 5 divisions to the left of the zero, and at the beat after it, 20 to the right of the zero, we should know that the passage had occurred $\frac{5}{25}$ or $\frac{1}{5}$ of a second after the first beat.

The observer must choose an odd number of half-vibrations, such that the interval consists of from 12 to 20 seconds, giving time enough to look at the clock and get a fresh start for counting by ear. The number of half-vibrations chosen must be odd, in order that alternate passages to right and left may be observed.

Suppose, for instance, it is decided to observe every 7th passage. It is not necessary to count all the 7th passages, but the approximate time of vibration being known the time at which, say 11, 7th passages may be expected, starting from an even minute, should be written down and placed conspicuously in front of the observer. This is merely a **convenient plan to** ensure the right passage being read in each **case.**

The observer then waits till a passage occurs at or near an **even** minute, and begins to observe alternately the passages in **the two** directions, which occur most nearly at the times written in the list before him.

For convenience **of writing** down, passages in one direction chosen arbitrarily **are called +**, and those in the other direction **—**.

They are called **out and** written down by an assistant in two columns, the + passages on one **side and the —** passages on the other.

For **a reason which** will **be seen immediately the set must con-clude with a** passage **of the** same sign **as** that with which it **began, so** that if, for instance, **we** begin with a **+** passage and **take 5 —** passages there will **be 6 +** passages. Suppose, **now,** there are 6 + and 5 — readings, the mean of all the — readings should be obviously equal to the middle reading. The object of having 5 is to ensure greater accuracy. The mean of the 5 negative readings is written **down as**

  'Time of mean middle negative passage from — observa-
      tions,   h.   m.   s.'

But also the mean of the 6 + observations should give the **time of** middle *negative* **passage; for** if they are divided into pairs, **1 : 6, 2 : 5, 3 : 4, the first** member of each pair is as much **before the middle —** passage as the second is after it.

Their **mean is then written down,**

  'Time of mean middle negative passage from + observa-
      tions,   h.   m.   s.'

The agreement of these two means is a test of **the** goodness of the observation. Their mean is written down,

  'True time of middle negative passage,   h.   **m.   s.'**

The magnet is **now left to** itself for **a** period **of** between 200 **and 300** half-vibrations, and then a second "true time of mean **middle negative passage"** is determined by 11 **observations, 6 +** and 5 —, exactly similar to the first set. The interval **between** the two middle — passages **is now** determined by sub-**tracting one time** from the other, and written down,

  '**Interval between** mean middle negative passage of set I.
      and ditto of set II.,   m.   **s.'**

The time of a half-vibration is again determined by taking the means of several intervals, taken at random in either set; or better, picked out from those observations in which the observer feels most confidence. This is written down,

'Time of half-vibration, 2nd approximation,   seconds.'

It is obvious that there must have been an *even whole number* of passages between the middle negative passages of sets I. and II.

The interval between them is divided by the 2nd approximation to the time of a half-vibration. The nearest even whole number to the quotient is the number of half-vibrations between the mean middle negative passages of sets I. and II.

This interval being then divided by this even whole number, the quotient is the 3rd approximation to the time of a half-vibration and is written down,

'True actual time of half-vibration =   seconds.'*

This, however, is not quite what is wanted; we do not want to know the actual time of a vibration, but what the time would have been if all disturbing causes had been removed, and the magnet had been at the standard temperature.†

The temperature must be observed at the beginning and end of each set. In Plate XIII. a thermometer C is shown attached to the magnet box.

CORRECTIONS.

In the vibration observations a correction has to be applied for—

Clock rate.—If the clock is gaining or losing the beats are not exactly seconds; the correction to be applied is, if s be the number of seconds lost or gained in a day,

$$\text{True time} = \text{obs. time} \times \frac{86400 - s}{86400}$$

where s is + for gain, — for loss. 86,400 is the number of seconds in a day.

In the deflection observations corrections have to be applied for—

Expansion, and error of graduation of bar D.

Distribution of magnetism on suspended and deflecting magnets.

* To make this line a correct statement for " seconds," read "clock beats."

† It is of no importance what temperature is chosen as the standard, as long as it is the same for the whole series of observations.

Alteration of the same by mutual induction of the magnets.
Variation of distance and direction with angle of deflection.
In both observations corrections have to be applied for—

Temperature.—The effect of temperature on the moment of the
magnet is determined by separate observations, and the diminution for each degree through which the temperature increases is
observed. The correction is **not** constant at all temperatures. A
formula, which has been **found by experiment to** be approximately **true,** is as follows :—

If t be the observed **temperature, and** $t_{\circ}$ the adopted standard
temperature,

$$\text{Magnetic moment at } t_{\circ} = (\text{mag. mom. at } t) \left[q\,(t_{\circ} - t) + q'\,(t_{\circ} - t)^2 \right].$$

and q' and q **are numbers which are constant for the same** magnet, **but** different **for different magnets. They must both be**
determined separately for **each** magnet **by experiment.**

Any *decrease* in the **magnetic** moment causes a corresponding
increase in the *square* **of** the **time** of vibration, so the above correction **must** be applied **to the** square of the time of vibration in
the inverse direction.

Torsion.—The **torsional force of** the suspending thread decreases the time of vibration, and has to be corrected for.

The torsion head being turned 90° alternately in the two directions, the mean of the deflections produced is called u.

The magnetic directive force is Hm, and if T be the force of
torsion the ratio of the force of torsion to the magnetic force is

$$\frac{\text{T}}{\text{H}\,m} = \frac{u}{90^\circ - u}$$

where u equals the angle through which **the magnet is deflected**
by a twist of 90° **in** the thread.

Also, **a** correction has **to be applied for the** moment of inertia and arc of vibration **of the** magnet.

Specimen Observations.

As a specimen of these **kinds of observations, the author** here
inserts some extracts from **the** details **of a comparison** of magnetic force at his laboratory, and at Kew.*

The same magnet was vibrated at Kew Observatory and at
the author's laboratory at Pixholme, Dorking.

From some preliminary experiments the approximate time of a

---

* Phil. Trans., 1877, page 22.

half-vibration at Pixholme was found to be 3·693 sec. Five different double sets of observations were taken.

The following are the details of set (1), Parts I. and II. Every seventh passage was observed :—

PART I.
April 23, 1876, P.M.—Mean temp. 13·9° C.

Passages in (+) direction.			In (−) direction.		
hr.	m.	sec.	hr.	m.	sec.
6	22	3½			
			6	22	29
6	22	55			
			6	23	20½
6	23	46½			
			6	24	12½
6	24	38¼			
			6	25	4
6	25	30			
			6	25	56
6	26	21½			
Mean 6	24	12·458	6	24	12·400

Time of middle negative passage.

				hr.	m.	sec.
From (+) obs.	.	.	.	6	24	12·458
From (−) obs.	.	.	.	6	24	12·400
		Mean		6	24	12·429

PART II.

Passages in (+) direction.			In (−) direction.		
hr.	m.	sec.	hr.	m.	sec.
6	39	2			
			6	39	28
6	39	53½			
			6	40	19½
6	40	45¼			
			6	41	11½
6	41	37			
			6	42	3
6	42	28½			
			6	42	54½
6	43	20¼			
Mean 6	41	11·83	6	41	11·26

Time of middle negative passage.

				hr.	m.	sec.
From (+) obs.	.	.	.	6	41	11·83
From (−) obs.	.	.	.	6	41	11·26
		Mean		6	41	11·545

Interval between mean negative passage I.
and „ „ II.
16 min. 59·251 sec. = 1019·251 sec.

If we divide this by 3·693, the approximate period of a half-vibration, the nearest even whole number to the quotient will be the number of passages in the time. Dividing by 3·693 we obtain the quotient 275·99 which gives 276 half-vibrations in the time.

Dividing 1019·25 sec. by 276 we obtain for the uncorrected time of one half-vibration 3·6929 sec.

This and the other four sets being corrected gave the times of vibration at Pixholme on certain dates. The times of vibration at Kew were determined on certain other dates. Also the ratio between the magnetic force at Kew, on the dates of the Pixholme and on the dates of the Kew observations is known from the records of the bifilar.*

We assume that the ratio of the magnetic force at Pixholme and Kew at the same time is constant.

Then we have

$$\text{Ratio at any time of } \left\{ \begin{array}{l} \text{Hor. Force at Pixholme.} \\ \hline \text{Hor. Force at Kew.} \end{array} \right.$$

$$= \frac{1}{\begin{array}{c} \text{Sq. of vibr. time} \\ \text{at Pixholme} \end{array}} \left\{ \begin{array}{c} \text{(Sq. of vibr.} \\ \text{time at Kew)} \end{array} \right. \left\{ \begin{array}{c} \text{Hor. Force at Kew at time} \\ \text{of Kew vibrations.} \\ \hline \text{Hor. Force at Kew at time} \\ \text{of Pixholme vibrations.} \end{array} \right.$$

$$= \frac{(\text{Vibr. time at Kew})^2}{(\text{Vibr. time at Pixholme})^2} \cdot \frac{\text{Hor. Force at Kew at time of Kew vibr.}}{\text{Hor. Force at Kew at time of Pixholme vibr.}}$$

We see that if the same magnet could have been vibrated at the same time at Kew and Pixholme, the ratio would have been expressed by the first term only of this product.

A magnet of the size usually supplied will swing for nearly an hour between the times when the arcs of vibration are too large and too small for observation.

OBSERVATIONS OF DECLINATION WITH THE UNIFILAR MAGNETO-METER.

For determining the declination, that is, the angle between the astronomical and magnetic meridians of a place at any time, the same arrangement (Plate XIII.) of the instrument is used as for vibration observations. N is a little plane mirror called the transit mirror, by reflection in which the sun can be seen in the telescope. "There† are three adjustments required for the transit mirror.

* Page 190.
† Admiralty Instructions for Magnetic Surveys, p. 24.

"1st. The axle to which the mirror is attached must be horizontal. This adjustment is performed by means of a riding-level.

"2nd. The mirror must be parallel to the axis of the cylindrical axle to which it is attached. This adjustment is made by means of a screw at the back of the mirror, as follows :—Turn the circle so that any well-defined object sufficiently elevated can be reflected into the telescope. Bisect the object by the wire of the telescope; reverse the axis, and observe whether the object remains bisected by the wire; if not, by the adjusting screw alter the inclination of the mirror until it is half the distance from the wire. Reverse again and again, until the object remains bisected before and after reversal of the axis.

"3rd. The line of collimation of the telescope must be perpendicular to the axis. Having made the first two adjustments, this adjustment may be made thus :—Suspend a plumb-line of some length in a sheltered position, or, if possible, within a house (the weight should swing in water to prevent oscillation). Turn the circle until the wire bisects the plumb-line, as seen directly; read the circle and turn it through exactly 180°. Observe whether the upper part of the plumb-line, when reflected into the telescope, coincides with the wire; if not, the adjusting screws must be moved until it does. In this operation it will be necessary to remove the magnet-box and suspension-tube. When this adjustment is completed, the adjusting screws ought to be fixed as tightly as possible.

"In the instruments most recently constructed, the telescope is furnished with a collimating eye-piece, by which, when the plane of the transit-mirror is vertical, the image of the wire of the telescope will be seen by reflection from it. By means of the proper adjusting screws, both the second and third adjustments may be effected by making the wire seen directly, coincide with its image seen by reflection before and after reversal of the transit axis. Both these adjustments can thus be readily verified before each observation."

DETERMINATION OF THE ASTRONOMICAL MERIDIAN.

In using the instrument—

1st. The magnet must be raised by the rack-work, so that the transit mirror can be seen in the telescope through the glass windows in the end of the box.

N

Then the whole upper part of the instrument is turned " in azimuth " (that is, round the vertical axis) and the transit mirror "in altitude" (that is, round its horizontal axis) till the sun is seen in the telescope just to the east of the cross wires. The circle is now clamped and the times at which both limbs of the sun pass the cross wires are noted with a chronometer. The verniers are read by means of the microscopes. Then, to eliminate any error in the adjustment of the transit mirror, it must be reversed in its bearings and the observations of the sun repeated and the time again noted.

From these observations and a knowledge of the time at the place, the latitude and the approximate longitude, the direction

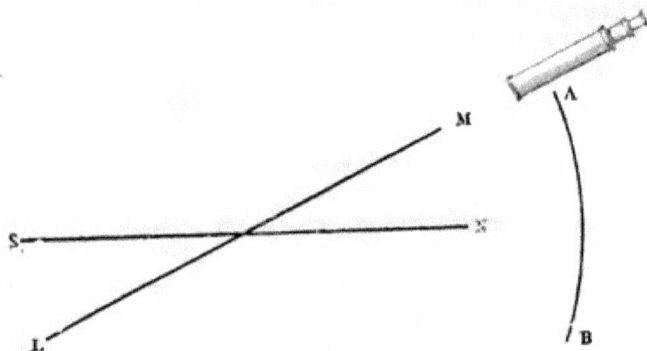

Fig. 56.

of the astronomical meridian can be found by trigonometrical calculation.*

DETERMINATION OF THE MAGNETIC MERIDIAN.

The magnet is now lowered and observed by means of the telescope, the circle being turned till the telescope is approximately in the line of the magnetic meridian. The circle is then clamped and turned by means of the tangent screw until the

* Formula of reduction :—

Let a = Polar distance of Sun at the time,
b = Co-latitude of the place of observation,
C = Hour angle of Sun at the time,
A = Azimuth of Sun from south,
B = Any angle.

Then $\tan \frac{1}{2}(A + B) = \dfrac{\cos \frac{1}{2}(a - b)}{\cos \frac{1}{2}(a + b)} \cot \frac{1}{2} C.$

$\tan \frac{1}{2}(A - B) = \dfrac{\sin \frac{1}{2}(a - b)}{\sin \frac{1}{2}(a + b)} \cot \frac{1}{2} C.$

$\frac{1}{2}(A + B) + \frac{1}{2}(A - B) = A.$

centre division of the scale in the magnet comes upon the spider line. The verniers are now read. The magnet is then inverted, that is, it is turned through 180° round the line joining its poles. The circle is again adjusted till the spider-line comes on the centre of the scale, and the verniers read. The object of this double reading is this:—It is impossible to make sure that the magnetic axis of the magnet coincides with the centre division of the scale. There will usually be some small angle between them. The mean of "erect" and "inverted" readings of the scale gives the magnetic meridian independent of the magnitude of this angle. For in fig. 56 let S N be the magnetic meridian

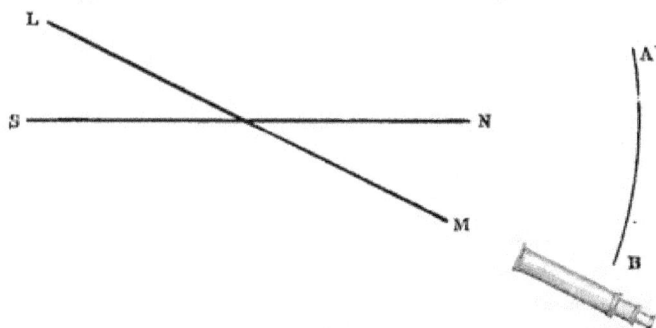

Fig. 57.

. and L M the direction of the middle division of the scale when the magnet scale is erect; then A will be the position of the telescope.*

Now let the magnet be inverted. The direction S N will remain unaltered but the direction L M will make an angle with it in the opposite direction and equal in magnitude to the former angle, and the telescope will have to be moved to the position B, fig. 57. The mean between the two positions A, B, of the telescope is the true direction S N. It can be easily seen that half the difference between the readings A and B is the angle between S N and L M, when this is known it will save inverting the magnet when observations have to be made quickly.

"The† torsion of the thread should be removed at every pos-

* The angle between L M and S N has been much exaggerated for clearness in the figures. In no actual instrument would it be more than a small fraction of a degree.

† Admiralty Instructions for Magnetic Surveys, p. 23.

N 2

sible opportunity. This is done by removing the magnet and substituting a brass bar of *equal weight,* allowing the bar to hang until it has assumed a steady position, and turning the top of the suspension-tube until the bar hangs steadily in the line of the telescope. The magnet may then be replaced for observation, the scale being always made horizontal and the divisions erect. In replacing the magnet care should be taken that a turn or half-turn of torsion is not introduced into the thread. Whenever time allows, the torsion should always be removed."

To bring a Swinging Magnet to rest.

In all magnetic observations it is almost necessary to be able to bring a swinging magnet to rest quickly. After a little practice this is easily done by the use of a small magnet held in the hand and suddenly moved to or from one end of the swinging magnet.

If we hold the hand magnet so that it repels the near end of the swinging magnet, it should be brought suddenly up to the swinging magnet as the latter is moving towards the observer, and just before it passes the zero point. The effect will be to first stop the swinging magnet and then give it an impulse in the opposite direction to that in which it was moving; but by removing the hand magnet the instant that the swinging magnet has stopped, the observer will, after a little practice, be enabled to stop a magnet almost dead at the zero point. It is not necessary to carry a special magnet for the purpose, as the steel lever or screw-driver belonging to the instrument, if magnetized, answers very well.

Barrow's Circle—Inclination and Total Force.
Observations of the Inclination with Barrow's Circle.

Plate XIV.

Barrow's Circle.—As arranged for determining the Inclination, this instrument is shown in Plate XIV. The outline of the construction is as follows :—

The whole upper part of the instrument turns on a vertical pivot over the circle E.

The needle C moves in a vertical plane on a horizontal axis.

An inch or two behind it is a sheet of ground glass, B, and two or three inches in front of it is the circle G, on which turns an

PLATE XIV.—BARROW'S CIRCLE, ARRANGED FOR INCLINATION.

arm moved by a tangent screw, P, and carrying microscopes, DD,
and verniers, II H. The object of having the arm so far in front
of the needle is to prevent the latter being affected by specks of
iron which may be present as impurities in the brass.

To observe the position of the needle the microscope arm is
turned until the ends of the needle are seen in the microscopes.
The arm being then clamped is adjusted by P, the tangent screw,
until one end of the needle is exactly in the centre of the field
of one of the microscopes. The verniers are then read, and a
similar observation made of the other end. The lenses outside
the microscopes are for reading the verniers.

To set the plane of the circle in the magnetic meridian we
must remember that when this plane is at right angles to the
magnetic meridian the whole of the horizontal component acts
in pressing the pivots of the needle against their supports, and
only the vertical component tends to turn it on its pivots.

The needle will therefore stand vertical when the plane of the
circle is at right angles to the magnetic meridian. The microscope arm being placed vertically, one of the verniers, say the
bottom one, is adjusted to 90°, and the circle turned round on its
vertical pivot till the end of the needle is seen in the centre of
the field. The horizontal circle is then read.

The top vernier of the microscope arm is now adjusted to 90°,*
and the horizontal circle again adjusted till the top of the needle
is seen in the microscope. The horizontal circle having been
read, the needle is reversed in its bearings; that is, the end of
the pivot, which formerly pointed to the front of the circle, now
points away from it, and both observations are repeated.

The mean of the four readings of the horizontal circle is the
position of the vernier when the plane of the vertical circle *is at
right angles to the magnetic meridian.* On the circle being
turned round on its vertical pivot and set so that the reading of
the vernier on the horizontal circle differs by 90° from its former
mean value, the plane of the circle comes into the magnetic
meridian.

The pivot of the needle, when in use, simply rolls on plates of
agate. To ensure its being in the centre it can be lifted off the
agate by two metal **Y**s worked up and down by means of the

---

* If the instrument were perfect, both verniers would read 90° at once. but
in practice there is always a difference of 1' or 2'.

side handle Q. When lifted in these the pivot falls to the centre, and when they are lowered it is deposited in the right position on the agates. If it turns it rolls out of this position; but, after being raised and lowered a few times, it falls into the right position while in the right direction.

The handle Q having been sufficiently worked, the microscope arm is adjusted so that the two ends of the needle are respectively observed, and the corresponding circle readings noted by the verniers.

The process is now repeated; if the north end of the needle was read first in the previous observation, the south end is read first now, and *vice versâ*.

This gives us four readings.

The vertical circle is now turned on its pivot through 180°, as indicated on the horizontal circle, so that if the front of the circle was formerly towards the east, it is now towards the west.

Four similar readings are taken.

The magnet is now lifted from its bearings and reversed, so that the end of the pivot, which formerly pointed to the front of the circle, points away from it.

The whole eight readings before described are then repeated.

The magnet is then taken out and put into the wooden block K and secured by means of the brass catch L, and its magnetic polarity is reversed by drawing two bar magnets over it from the centre outwards.*

The use of this process is to eliminate any inequality in the balance of the needle or its pivot.

The reader should remember that the intensity of the magnetization does not affect the process of determining the dip, and a slight difference between the direction of the magnetic axis and the line joining the points of the needle is eliminated by reversing the pivot on its bearings.†

The whole sixteen observations are now repeated with the re-magnetized needle.

The mean of the thirty-two readings is the direction of the dip.

Before commencing to observe, the instrument must be accurately levelled by means of the spirit level R and the screw feet A,

---

* See fig. 46, p. 147.
† Compare p. 170.

PLATE XV.—BARROW'S CIRCLE, ARRANGED FOR TOTAL FORCE.

so that when the verniers read zero, a needle seen in the micro-scopes would be exactly horizontal.

It is usual to take two such sets of 32 readings, and take the mean; but if the results of the two sets differ by more than 3' or 4', another must be taken.

OBSERVATIONS OF TOTAL FORCE WITH BARROW'S CIRCLE.
Plate XV.

Plate XV. shows the instrument arranged for observations of total force. The method used is due to Dr. Lloyd.

" For * this purpose the instrument is furnished with two addi-tional needles, which may be called, for distinction, Nos. 3 and 4, *the poles of which are at no time to be reversed or disturbed ;* Nos. 1 and 2 being needles used for observing the inclination in the way just described. No. 3 is an ordinary dipping needle ; No. 4 is a similar needle loaded with a small fixed and constant weight, acting in opposition to magnetism. The frame, carrying the microscopes of the circle, is also fitted to receive and to retain No. 4 securely in a constant position, when it is used as a deflector of No. 3.

" The observations consist of two processes ; by the one process the " position of equilibrium " † is observed of No. 3 between the action of the earth's magnetism, and that of No. 4 used as a deflector, having its North pole directed alternately towards the magnetic North and South ; and by the other process the position of equilibrium of No. 4 is observed between the action of the earth's magnetism and that of the small constant weight with which it is loaded."

The first process gives us the relation between the earth's total force and the magnetic moment of magnet No. 4.

The second process gives us the relation between that magnetic moment and the moment of a known weight placed at a known distance from the axis.

The product of these two ratios is the ratio of the earth's force to the known moment of the weight ; that is, it gives us the earth's force in absolute measure.

* Admiralty Instructions for Magnetic Surveys, p. **27**.

† The "position of equilibrium" is the position of the needle when it is exactly at right angles to the microscope arm D D and deflecting magnet. In this position the ends of the needle are seen in the microscopes.

Thus—

$$\text{Ratio of } \frac{\text{Earth's mag.}}{\text{Mag. mom. of No. 4}} = A$$

a known quantity.

$$\text{Ratio of } \frac{\text{mag. mom. of No. 4}}{\text{moment of known weight}} = B$$

a known quantity.

Multiplying and cancelling, we have

$$\text{Ratio } \frac{\text{Earth's mag.}}{\text{mom. of weight.}} = A B, \text{ a known quantity.}$$

OBSERVATIONS AT SEA.—THE FOX CIRCLE.—Plate XVI.

None of the instruments hitherto described are suitable for use at sea. The motion of a ship would make them perfectly useless. As the greater part of the magnetic surveys are made at sea, it is necessary to have an apparatus which can be used on board ship. Such an apparatus is the instrument invented by the late Mr. R. W. Fox, which, while sacrificing only a little accuracy, can be used under almost any circumstances.

" The [*] instrument consists of two graduated circular rims, Plate XVI., whose planes are vertical and perpendicular to the line joining their centres. The graduations in each are to 15', the zero points being in the horizontal diameter.[†] The aperture of the inner rim F is less than that of the outer one E, so that the divisions of each can be seen simultaneously by an observer in front, and should exactly correspond.

" The needle B swings between these two rims, but much nearer to the inner than the outer one. Its axis, which should be in the line joining the centres of the two graduations, is terminated by very short cylindrical pivots, which work in jewelled holes. This axis carries a small grooved wheel H, round which passes a thread of unspun silk, furnished with hooks G G, for hanging weights on in taking intensity observations. The whole is enclosed in a brass cylindrical box. This box stands on an azimuth base, in which turns a vernier plate, as in a common theodolite.

* Walker, *Terrestrial and Cosmical Magnetism,* p. 218.

† By an error in Plate XVI. the zeros are shown displaced. They should be in a horizontal line, and the 90°s. in a vertical line.

PLATE XVI.—THE FOX CIRCLE.

"The azimuth plate is fixed to another called the foundation plate, which stands on foot-screws. Sometimes the instrument is made so that it can be screwed on to a stand like that of a theodolite. In the box is placed a thermometer for noting the temperature at the time of an observation.

"METHOD OF USING THE INSTRUMENT.

"The object of having two series of coincident graduations, one on each side of the plane in which the needle swings is two-fold. In the first place they prevent any error of parallax in reading off, as the eye is brought into the line joining the corresponding divisions; and in the second place, the divisions of the outer circle serve as a vernier.*

"The instrument may be used for determining *dip* or *intensity.*"

To prevent the needle "sticking," the end of its bearing is gently rubbed by means of the rubber C (Plate XVI.), which consists of a flat piece of horn with a number of saw cuts

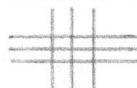

to roughen the surface.

"DIP.†

"The plane of the magnetic meridian having been determined, the face of the instrument is made to coincide with it, and both ends of the needle read. The box is then turned through $180°$

---

* "For if the outer circle be n times further from the plane of the needle's motion than the inner one, then the line of sight which passes through the point of the needle must move over n divisions of the former to make the needle appear to move over one division of the latter.

"Thus suppose the graduations of the two circles to be to p, then each division on the outer rim will correspond to $\left(\dfrac{p}{n}\right)$ on the inner rim. The method, therefore, is to read off the nearest division to the end of the needle, and then to carry the eye along the outer rim till this division is in the same straight line with the eye and the point of the needle. If m be the number of divisions passed over on the outer rim $\dfrac{m\,p}{n}$ is the number of minutes to be added to the previous reading.

† Walker, *Terrestrial and Cosmical Magnetism,* p. 220.

in azimuth, and the observations repeated. The mean of the whole will indicate the dip approximately.

" This approximate value is corrected thus :—

" There is a small magnet (called the deflector) K inclosed in a brass tube fitted with a screw. This tube is now screwed on to the back of the instrument, so as to repel the end of the needle nearest it, and adjusted at a given angle from the observed dip.

" When the needle has come to rest both ends are read. The deflector is then transferred to an equal distance on the other side of the dip, and the two ends of the needle read as before. This operation is repeated with the face of the instrument turned through 180°. The mean of the four readings is the corrected value of the dip.

" Most instruments of this construction are fitted with two deflectors, one to repel the north and the other the south end of the needle. Mr. Fox, however, appears to prefer the use of only one deflector, as more consistent results are usually obtained with small than with large angles. Sometimes, as was the case with the instrument on board the *Erebus,* in the Antarctic Expedition of 1839-43, it is necessary to use both deflectors at once, in consequence of the weakness of the deflecting force when they are used separately.

" INTENSITY.

" The deflector or deflectors being removed, the silk thread is placed over the grooved wheel, and a given weight suspended from one of the hooks, and when the needle has come to rest each end is read. The weight is removed and suspended from the other end of the string, and both ends of the needle read as before. Half the difference between the two readings will be the deflection." *

This gives us the ratio of the known moment of the weight to the product of the earth's force into the moment of the magnet.

To determine the moment of the magnet it must be dismounted, put into the brass box K, and used as a deflector, while another needle without weights is substituted for it in the centres.

This gives the ratio of the magnetic moment of the first magnet to the earth's force.

As we now know both the ratio and product of the earth's force

* Walker, *Terrestrial and Cosmical Magnetism,* p. 221.

and the moment of the first magnet, the former quantity can be determined by a similar calculation to that given for the deflection observations with the unifilar (p. 167).

This last method of observation is an adaptation of Dr. Lloyd's method (described p. 183), made by Mr. Welsh, of Kew Observatory.[*]

" The[†] correction for temperature is obtained by placing the instrument under a glass receiver, and admitting heated air under it. The needle being deflected at a given angle from the dip by a given weight, the change in this angle, corresponding to an increase of temperature, must be noted. The *actual* change will depend on the magnetism of the needle; but the *ratio* appears to be very nearly uniform within the limits of the range of temperature in this climate."

The following is the testimony of Captain Ross to the merits of Mr. Fox's instrument:—

" By means of the admirable contrivance of Mr. R. W. Fox we are able, in tolerably moderate weather, to determine the three magnetic elements with even more precision on board our ships than they are susceptible of on shore, on account of the unknown and indeterminable amount of local attraction; and even in the heaviest gales, after a little practice with his instrument, they may be observed with sufficient exactness to afford very useful and important information. Throughout the whole distance of between three and four thousand miles from Kerguelen Island to Van Diemen's Land, we could not have derived a single satisfactory result with the instruments in common use; and this portion of the ocean, at least, must for the present have remained a blank upon our charts. But with Mr. Fox's apparatus, the dip and intensity observations were accomplished in an almost uninterrupted series of daily experiments."— *Ross's Antarctic Voyage*, vol. i. chap. v.

" The principal cause of this superiority of Fox's instrument in taking observations at sea is its stability, arising from the mode in which the needle is hung. As we have seen, the axis terminates in very short cylindrical pivots, which work in jewelled holes. By this construction any displacement of the needle in consequence of the rolling or pitching of the vessel is prevented;

---

[*] Instructions for Magnetic Surveys, p. 31.

[†] Walker, Ibid., p. 225.

whilst the loss of sensibility, which undoubtedly results from this mode of suspension, is a positive gain under such circumstances. Another advantage offered by the instrument is the substitution of the statical for the vibration method in observations on the intensity. With the delicate suspension required in vibration experiments, such an observation, except in very rare circumstances, would be hopelessly impossible."

CHAPTER XV.

SELF-RECORDING INSTRUMENTS.

Plates XVII. and XVIII.

For the study of the daily and hourly changes of the magnetic elements, self-recording instruments are necessary.

The principle of these instruments is, generally speaking, as follows :—

A mirror is attached to the moving magnet, and a spot of light from a lamp thrown by it on a piece of sensitized photographic paper moved continuously by clock-work.

If the magnet remains at rest, a straight line is traced ; if it moves, a zigzag line.

Three elements are observed, viz. :—

Declination.
Horizontal force.
Vertical force.

From the ratio of the last two of these the inclination can be calculated, so it is not observed separately.

DECLINATION.

Plates XVII. and XVIII. (fig. 1 in each).

The portion of the apparatus which records the declination consists of a suspended magnet with a mirror attached ; the magnet being hung by a single fibre as in the Unifilar magnetometer. It stands on a massive stone pier. Light from a gas-lamp passes in through a slit and collimating lens, falls upon the mirror, and is there reflected inside a wooden tube into the box, fig. 4, Plate XVII., which contains the barrels and clock-work shown in fig. 4, Plate XVIII. The light falls on one of the horizontal barrels. Round this barrel is fastened a sheet of sensitized paper. If the barrel were at rest and the magnet moved, a black line would be traced on the barrel parallel to its axis and of a length corresponding to the extreme motions of the magnet. If the magnet remained at rest and the barrel revolved, a line would be drawn round the barrel, which, when the paper was unrolled, would be a straight line perpendicular to its axis.   When however the clock moves the barrel **steadily round, and** the magnet oscillates at the same time, an irregular curved or zigzag line is traced round the **barrel,** the distance of **any part of** which **from** the base line gives the direction **of the magnetic** meridian at **the** corresponding time.

The base line is drawn by a spot of light reflected from a fixed mirror **whose direction with** regard to the astronomical meridian is known.

The mirrors are made in two halves of a disc, the one below fixed to the base, the upper one to the suspended magnet.

A screen moved by the clock cuts off the light from the fixed mirror for a minute or two, commencing at the beginning of every other hour.   This gives a time scale, so that even if the clock does not go correctly, different diagrams can be compared as long **as the rate** is known.

Horizontal Force.
Plates XVII. and XVIII. (fig. 2 in each).

This is registered by means of a bifilar suspension.

The two ends of a fine steel wire are attached to a horizontal **screw whose direction is** at right angles to the magnetic meridian. The **magnet is suspended by a** pulley round which the wire passes.   **Thus the** magnet is practically suspended by two wires always **equally tight.**

The magnetic force **tends** to turn the magnet round.   If it turns it has to commence to twist the wires and raise itself. A fraction of the weight **of** the magnet then acts against the magnetic moment multiplied by the horizontal force; as the **latter** increases, the magnet turns further, till, **by** its change of position, **a** greater proportion of its weight acts in balancing the magnetic **force.**   The position of **the magnet** then indicates the strength **of the** horizontal force.[*]

The value of the horizontal force indicated by a given position **is determined by vibration** experiments, the results of which are **compared with the** mean position of the bifilar magnet during

* **The whole deflection being very small.**

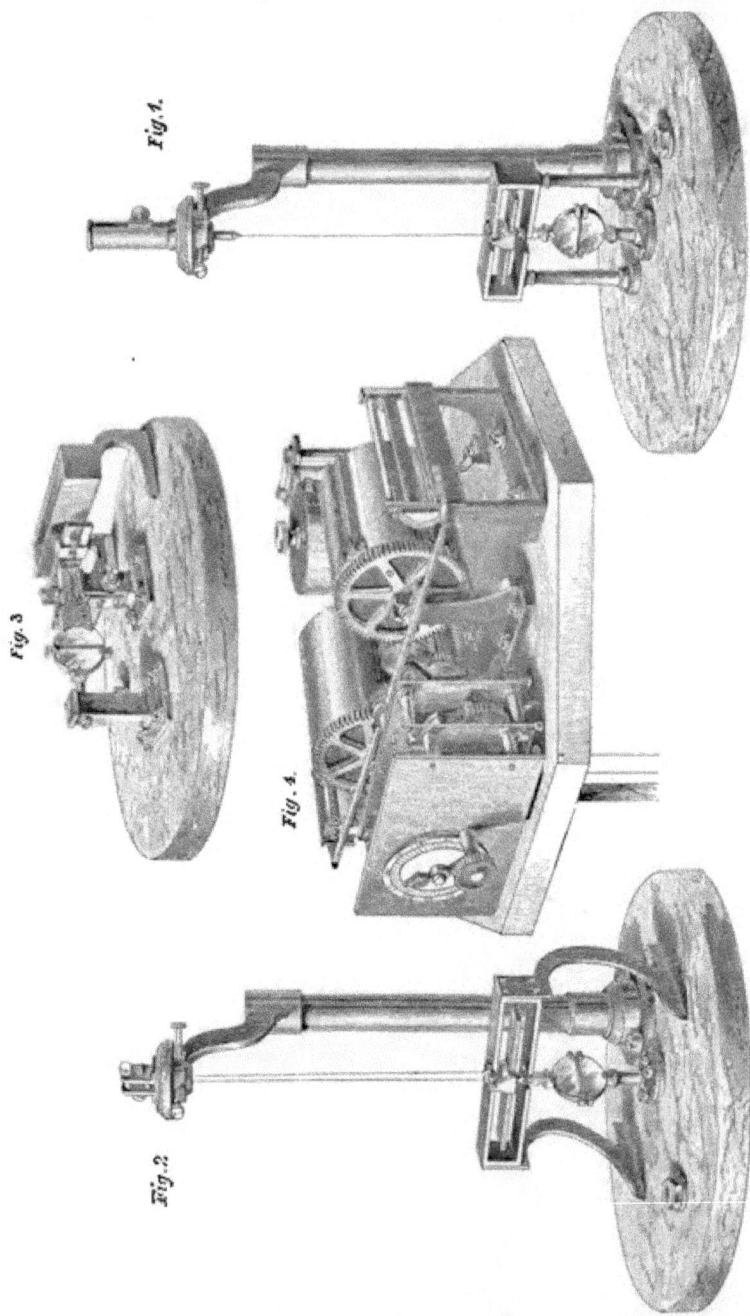

Fig. 1.

Fig. 3.

Fig. 4.

Fig. 2.

PLATE XVIII.—KEW SELF-RECORDING MAGNETOMETERS. DETAILS.

the time which the vibration experiments last. The motions are registered on the other horizontal drum in precisely the same manner as those of the declination magnet.

For *small* variations of horizontal force the distances from the curve to the base line may be considered to be proportional to the force. The magnet can be raised and lowered for adjustment by turning the screw.*

VERTICAL FORCE.
Plates XVII. and XVIII. (fig. 3 in each).

A steel bar is mounted on a pivot provided with agate knife edges, and accurately balanced on agate planes. It is then magnetized and the marked end dips. It is brought back to a nearly horizontal position by a brass weight fixed near the unmarked end, and so adjusted that the centre of gravity of the whole system is a little below the point of the support. The effect of this is that the more the magnet is displaced from the horizontal position the greater is the moment of the weight tending to bring it back.

Thus, if the vertical magnetic force decreases, the magnet moves so as to become more nearly horizontal; if it increases, the magnet turns in the other direction. Its position at any time is a measure of the intensity of the vertical force.

Its motions are registered on a drum in the same manner as the declination and horizontal force, the only difference being that as in this case the oscillations of the spot of light are in a vertical line, so the drum must turn on a vertical instead of a horizontal axis.

DETAILS.

To avoid disturbance by currents of air all three magnets work in vacuo. The brass boxes being ground flat, glass receivers with ground edges are placed over them, and the air is pumped out by an ordinary air pump.

Each magnet is furnished with a telescope and scale, so that its position can be at any time observed by the eye.

---

* In some instruments, but not in those in use at Kew, it is arranged that as the screw turns it moves longitudinally through its bearings, but the wire or thread, as it winds in the screw thread, moves in the opposite direction at the same rate, so that the magnet only moves vertically. The use of a screw keeps the threads always the same distance apart, and prevents the thread winding on itself as it would on a plain bar.

The measured values of the curves have to be corrected for temperature. Therefore the variations of **temperature** are continuously recorded by fixing a thermometer in a slit in a screen, on one side of which is a light and on the other a drum covered with sensitive paper. Light passes through the upper part of the tube but is stopped by the mercury column. A curve showing the variations of the height of the column is thus traced, the whole of the paper on one side of it being white and on the other black.*

For economy two days' tracings of each of the magnetic elements are commonly taken on the same paper, the gas-burners being displaced slightly to one side at the end of the first day.

Plate **XIX.** shows the variations of the horizontal force on two consecutive days; the first day being one of ordinary change, the second being characterized by the occurrence of a violent magnetic storm. It is an exact reduced fac-simile of one of the photographic records. On two days of ordinary change the curves would not cross each other.

The original tracing is black on a white ground. In 1862, when it was taken, the arrangement for cutting off the light every two hours had not been introduced.

* In meteorological observatories, a similar method is also used for registering the variations of the barometer.

PLATE XIX.—KEW MAGNETOMETER RECORD.

193

CHAPTER XVI.

OBSERVATIONS ON TERRESTRIAL MAGNETISM.

SECULAR CHANGES.

THE DECLINATION.

THE following table shows the changes which have taken place in the declination since the year 1580 :—

Year.	Declination in London.	Year.	Declination in London.
1580	11° 17′ E.	1800	24° 36′ W.
1622	6° 12′ ,,	1806	24° 8′ ,,
1634	4° 5′ ,,	1815	24° 27′ ,,
1657	0° 0′ ,,	1820	24° 11′ ,,
1666	0° 34′ W.	1831	24° 0′ ,,
1672	2° 30′ ,,		At Kew.
1700	9° 40′ ,,	1858	21° 54′ W.
1720	13° 0′ ,,	1863	21° 13′ ,,
1746	16° 10′ ,,	1868	20° 33′ ,,
1760	19° 30′ ,,	1873	19° 58′ ,,
1774	22° 20′ ,,	1878	19° 14′ ,,
1790	23° 39′ ,,	1879	19° 6′ ,,
		1881	18° 51′ ,,

We see that the declination was easterly in London from 1580 to 1657. In 1657 it vanished, and the magnetic meridian coincided with the astronomical.

The needle then began to move to the west of north, and the westerly declination continued to increase till about 1815.

The needle then turned back, and has ever since been returning towards the astronomical meridian.

THE INCLINATION.

The following table shows the changes which have been observed in the inclination since its discovery in 1576 :—

O

Year.	Inclination in London.	Year.	Inclination in London.
1576	71° 50′	1838	69° 17′
1600	72° 0′	1854	68° 31′
1676	73° 30′		
1723	74° 42′		At Kew.
1780	72° 8′	1858	68° 23′
1790	71° 33′	1863	68° 12′
1800	70° 35′	1868	68° 2′
1818	70° 34′	1873	67° 52′
1821	70° 3′	1878	67° 43′
1828	69° 47′	1879	67° 42′
		1881	67° 41′

We see that from the time of its discovery to about 1723 the inclination at London increased. From that date to the present time it has decreased.

Force.

There is also a secular change in the horizontal and total force.

The Horizontal Force increases from year to year at a sensibly uniform rate. Its mean value at Greenwich was

$$\cdot1716 \text{ in the year } 1848,$$
$$\cdot1776 \text{ ,, ,, ,, } 1867,$$
$$(\text{at Kew}) \cdot1797 \text{ ,, ,, ,, } 1879,$$
$$\cdot1800 \text{ ,, ,, ,, } 1881.$$

The yearly increase is about ·00124 of the whole force.

The Total Force probably decreases from year to year. Its mean value at Greenwich was

$$\cdot4791 \text{ in } 1848,$$
$$\cdot4740 \text{ in } 1866,$$
$$(\text{at Kew}) \cdot4737 \text{ in } 1879,$$
$$\cdot4739 \text{ in } 1881.$$

Periodic Changes.

Magnetic Observations under the Direction of Major-General Sabine, *late* P.R.S.

General Sabine's work on Terrestrial Magnetism is contained in a series of papers, in nearly every number of the "Philosophical Transactions," from 1840 to the present time.

A great portion of his communications is taken up with a Magnetic Survey of the Globe, chiefly for the use of mariners. As however no general theory of the distribution of terrestrial magnetism can yet be said to be established, we will confine our attention to his observations and conclusions with regard to the variations of the several elements; and first we will say a few words on the methods used for separating disturbances of different periods.

Suppose we have observations of the elements for every hour for a long period, and suppose we want to know the change from day to day, we take means of all the hourly observations on each day, and the difference between the means of consecutive days will give the daily change.

If, however, we wish to find the mean hourly change, we take *the means of the observations made at the same hour on different days.* This gives us the hourly change and *eliminates the daily change.*

General Sabine's first general conclusion with regard to the secular change was, that there is a decennial period[*] in the larger disturbances of declination.

At Toronto and Hobarton the years 1843 and 1848 were periods of minimum and maximum disturbance respectively.

This decennial period " does not appear to connect itself with any of those divisions of time with which we are conversant, as depending on the relative circumstances of the sun, the earth, and her satellite."

Hofrath Schwabe has, however, made the remarkable discovery that the decennial period absolutely agrees, not only in its length but in the positions of its maxima and minima, with a period which he has discovered in the frequency and magnitude of solar spots.

Monthly means, taken in different years, have shown that there is an annual period of the disturbances corresponding to the apparent path of the sun in the ecliptic.

In southern stations the declination is easterly from May to September, and westerly in the remaining months of the year. In northern stations the reverse is the case.

The total force probably attains in Dublin a maximum in June and a minimum in February. In the southern hemisphere (Hobarton) the reverse is observed.

THE INCLINATION.

At Kew the inclination is below the mean in May, June, July,

* *Philosophical Transactions*, 1856, p. 361.

August, and above it in the remaining months of the year. At Hobarton the south inclination is below the mean from June to October inclusive, and above it for the rest of the year.[*]

DIURNAL INEQUALITY.

Hourly observations have shown that there is another period, coincident with the period of rotation of the earth on her axis.

Dr. Lloyd, speaking of the diurnal inequality in Dublin, says that its general features are as follows :—

" I. The easterly force diminishes from 7 a.m. or 8 a.m., and the north pole of the magnet moves *westward* until about 1 p.m., when the easterly force is a *minimum*.

" II. After 1 p.m. the easterly force increases and the north pole of the magnet returns *eastward*. This easterly movement continues until about 10 p.m., when the easterly force attains its greatest value.

" III. There is a second but much smaller oscillation during the night and early morning, the easterly force diminishing and the north pole moving slowly westward, for a few hours before and after midnight; after which it returns to the east until 7 a.m., when the easterly force is again a maximum.

" IV. In summer the westerly movement during the night disappears, the afternoon easterly movement continuing throughout the night, but at a slower rate. In winter, on the other hand, the morning easterly movement vanishes, and the magnet is almost in state of repose from 2 a.m. to 8 a.m.

" V. From the facts last mentioned it follows, that the greatest range in summer is that of the westerly movement from 7 a.m. to 1 p.m.; while in winter the greatest range is that of the easterly movement between 1 p.m. and 10 p.m."

LUNAR DIURNAL INEQUALITY.

A period has also been discovered corresponding to the lunar day.

Of it General Sabine says :[†] " The variation in each of the three elements constitutes a double progression in each lunar day; the declination has two easterly and two westerly maxima in the interval between two successive passages of the moon over

---

[*] Lloyd's *Treatise on Magnetism.*
[†] *Philosophical Transactions*, 1856, p. 505.

the astronomical meridian ; and the inclination and the total force have each two maxima and two minima due to the moon's action in the same interval, the variation passing in every case four times through zero in the lunar day. The easterly maxima of the horizontal deflection of the north-pointing end of the magnet synchronise with the moon's superior and inferior passages of the meridian ; the westerly maxima with the lunar hours of 6 and 18. The maxima of the increased magnetic force due to the moon's action occur about the lunar hours of 3 and 16, and the minima about 9 and 20."

Thus we see that the changes in the magnetic elements depend principally on the relative positions of the sun and moon with respect to the earth. Whether the greater part of the effect is due to direct magnetic action, or indirectly to the heating and cooling of the crust of the earth, we are not in a position to decide with absolute certainty.

Effect of Sunspots.

In 1859 a magnetic storm of unprecedented magnitude continued from August 28th to September 7th. Professor Balfour Stewart[*] has pointed out that this was synchronous with the period of maximum activity of one of the largest sun-spots ever observed.

The Aurora.

It is found that the appearance of certain kinds of auroræ coincides with certain periods of disturbance, but the observations at Point Barrow in 1852-3-4,[†] which are the best we yet have, cannot be said to have established any definite law of connection.

Broun's Observations.

On December 15, 1875, Mr. J. A. Broun, F.R.S., communicated to the Royal Society[‡] a paper " On the Variations of the Daily Mean Horizontal Force of the Earth's Magnetism produced by the Sun's Rotation and the Moon's Synodical and Tropical Revolutions," of which the following is an abstract :—

" The variations of daily mean horizontal force in the years

---

[*] *Philosophical Transactions*, 1861, p. 423.
[†] Ibid., 1857, p. 497.
[‡] Proc. Roy. Soc., xxiv., 1875-76, p. 231.

1844 and 1845 showed several well-marked oscillations, having periods of from 20 to 30 days, and amplitudes, in some cases, of more than one thousandth of the whole magnetic force. These oscillations were first attributed to lunar action ; afterwards they were found more probably due to the sun's rotation on his axis. The disappearance of these oscillations in the middle of well-marked series, their different amplitudes and periods, could not be explained except by the supposition that the solar action was not continuous, but only by fits periodic.

" The author was induced to believe lately that these differences in the oscillations were due to conjoint actions of the sun and moon ; he accordingly deduced the mean variations corresponding to three periods of 26, 29·5, and 27·3 days, the times of rotation of the sun derived from the magnetic observations, and of the moon's synodical and tropical revolutions respectively. He finds that the combinations of these three series of variations represent with considerable accuracy all the variations of the daily mean horizontal force of the earth's magnetism during each year ; so that the sun's rotation and the different positions of the moon relatively to the sun and the plane of the equator (or of the ecliptic) are found to produce all the differences in the amplitude and time, as well as the apparent disappearance of the oscillation.

" Cases of considerable and sudden diminution of the earth's magnetic force which happened in the years 1844 and 1845 are next examined ; and it is shown that these changes occur at intervals of 26 days, or multiples of 26 days; in one instance there are five successive recurrences at the exact interval of 26 days.

" As this period is that of the sun's rotation *relatively to the earth*, it appears to follow that the earth has some action on the sun, or (more probably) on some ray-like emanation from the sun, which causes these changes in the earth's magnetism.

" It is found also that these sudden variations occur more frequently when the moon is at a considerable distance from the equator and the ecliptic ; it would thus appear that our satellite has also an action on the cause of the great terrestrial magnetic disturbances."

CHAPTER XVII.

EARTH CURRENTS.

MAGNETIC observations are complicated by the existence of certain currents of electricity which move in the earth.*

Mr. C. V. Walker, F.R.S., superintendent of the South Eastern Railway telegraph lines, has made a series of investigations on these currents, and has communicated the result to the Royal Society, in 1861 and 1862.†

He thus summarizes his results :—

"The results arrived at in these two communications may be briefly summed up as follows:

" 1st. That currents of electricity are at all times moving in definite directions in the earth.

" 2nd. That their direction is not determined by local causes.

" 3rd. That there is no apparent difference, except in degree between the currents collected in times of great magnetic disturbance, and those collected during the ordinary calm periods.

" 4th. That the prevailing directions of earth currents or the currents of most frequent occurrence are approximately N.E. and S.W. respectively.‡

" 5th. That there is no marked difference in frequency, duration, or value between the N.E. and S.W. currents.

" 6th. That (at least during calm periods) there are definite currents of less frequency from some place in the S.E. and N.W. quadrants respectively.

" 7th. That the direction of the current in one part of a plane on the earth's surface (at least as far as the S.E. district of England is concerned) coincides with the direction in another part of

* See Chapter XX., " Action of Currents on Magnets."
† Proc. Roy. Soc., xi., 1860-62, p. 581.
‡ These would tend to set the magnets S.E. and N.W. respectively.

the plane; and if the direction changes in one part, it changes in all parts of the plane.

" 8th. That the relation in value between currents in a given part of the plane and currents in another given part is not constant, but is influenced by local meteorological conditions, and varies from time to time.

" 9th. That the value of a current of a given length moving in a given line of direction is not necessarily the same as that of a current of the same length on the same line of direction produced, and that their relative value depends on the physical character of the earth interposed between the respective points of observation, and is tolerably constant.

" 10th. That the currents which have formed the bases of these investigations are derived currents from true and proper earth currents and neither in whole nor in any appreciable part have been collected from the atmosphere, nor are due either in whole or in any appreciable part to polarization imparted to earth-plates by the previous passage of earth currents or of powerful telegraphic currents; nor are they due to any electromotive force in the earth-plates themselves.

" 11th. That the earth currents in question (at least the powerful currents present at all times of great magnetic disturbance) exercise a *direct* action upon magnetometers, just as artificial currents confined to a wire exercise a direct action upon a magnet.

PROFESSOR ADAMS' OBSERVATIONS.

At the meeting of the British Association in York in 1881, Professor W. G. Adams, F.R.S., read a paper on " Magnetic Disturbances and Earth Currents." [*]

Professor Adams shows by careful comparison of observations taken simultaneously in widely different localities, that they must be due to some cause external to the earth, and strong enough to affect its magnetism as a whole. He suggests that they may probably be due to tides caused by the sun and moon in the elastic solid crust of the earth and in the air. He considers that the tidal wave revolving round the earth may be regarded as if it was a conductor revolving round a magnet.

* Report, Brit. Assoc., 1881.

PART III.

ELECTRO-KINETICS.

A PHYSICAL TREATISE

ON

ELECTRICITY AND MAGNETISM.

Part III.

VOLTAIC ELECTRICITY AND
ELECTRO-MAGNETISM, OR "ELECTRO-KINETICS."

CHAPTER XVIII.

THE VOLTAIC BATTERY.

WE have said that if an insulated conductor be charged in any way whatever, in a very short time all parts of it will be found to be at the same potential. Suppose we take two conductors and charge them so that their potentials are different. If we now connect them by a wire, we shall find that after a very small fraction of a second the potential of the one will have diminished, and that of the other increased till they are at the same potential.

The one whose potential has diminished has lost a certain quantity of electricity, while that whose potential has increased has gained a certain quantity. Now, as no electricity has entered or left the system, the gain of electricity by the one must be equal to the loss by the other, and we are therefore justified in saying that electricity has travelled from one to the other along the wire.

This flow of electricity along the wire is called an "electric current," and takes place whenever a wire or other conductor is used to connect two conductors whose potentials are different.

When any two insulated charged bodies are connected, the flow which goes on until the potentials are equal, lasts but a very short time, as the potentials approach equality with great rapidity. If, however, instead of insulating the charged bodies, we connect them with a machine which, by the expenditure of work, will keep their potentials constantly different, the electric current will continue to flow along the connecting wire as long as the machine is in action.

The voltaic battery is such a machine. One view of the principle on which it is founded is this :—

If two metals be placed, near together but not in contact, in a liquid which acts chemically more upon one than upon the other, the metals become charged so that the one least acted on is of higher potential than the one most acted on. The difference of potential produced depends only on the nature of the metals and of the liquid, and not on the size or position of the plates.

As soon as the difference of potential has reached its constant value, the chemical action ceases.

If now the metals are connected by a wire outside the liquid, the difference of potential begins to diminish, and an electric current flows through the wire. As soon as the difference of potential becomes less than the maximum for the metals and liquid, chemical action recommences and brings it up to the maximum and thus, if no disturbing cause interferes, the current will continue till the metal most acted on is entirely dissolved.

This view of what takes place explains the action very well. It is not yet certain whether this is the true explanation, or whether we should say : On joining two metals either directly or by a wire, a difference of potential is observed. When the metals, still joined, are partly immersed in a liquid, which acts more upon one than upon the other, the chemical action equalizes the potentials, and in doing so causes a flow of electricity along the connecting wire. The moment the equalization of the potentials has commenced, the difference is renewed again at the point or points of contact between the metals; and so, if no disturbing cause interferes, a continuous flow of electricity is kept up till the metal most acted on is entirely dissolved.

The latter view has, in my opinion, more evidence to support it than the former. It will be more fully discussed in Chapter XIX., Contact Electricity.

When two metals are arranged as above described in a liquid, and are in metallic communication, the one which, if alone would be most acted on, entirely protects the other, and the arrangement is called a *voltaic circuit*, or cell.

In what follows we will call that part of the metal least acted on, which is not immersed in the liquid, the positive pole of the battery, and the corresponding part of the other the negative pole. If we bring the liquid in the battery to the potential of the earth, the plate at the higher potential will be positively charged, and that at the lower potential negatively charged, and the above convention will agree with our definitions in Part I.

In nearly all practical forms of the voltaic cell, the metal forming the negative pole is zinc; that forming the positive pole varies.

The simplest form of voltaic cell consists of a plate of copper and a plate of zinc (fig. 58) partially immersed in diluted* sulphuric acid, which acts on the zinc, but not on the copper. With such an arrangement, however, the current only continues for a very short time, and then ceases. Evidently some disturbing cause is acting. On examining the copper, it will be seen to be entirely coated with minute bubbles, which, if collected and tested, will be found to consist of pure hydrogen gas.

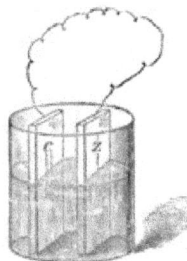

If a piece of zinc alone be dissolved in dilute sulphuric acid, the water is decom-

Fig. 58.

posed, and the oxygen combines with the zinc, and hydrogen is set free.†

When the decomposition occurs in a voltaic cell, the hydrogen is liberated, not at the surface of the zinc, but at that of the copper.

* Unless the contrary is stated, it is to be understood that "diluted sulphuric acid" means a mixture of 7 parts (by measure) of water with one part of acid (Coml. sp. gr. 1·845). In mixing, care must be taken to measure out the water first, and then to add the acid to it. It is very dangerous to add the water to the acid.

† The action is expressed by the following chemical equation :—

$$H_2 SO_4 + Zn = Zn SO_4 + H_2,$$

which expresses that one atom of zinc changes places with two atoms of hydrogen, converting the sulphuric acid into sulphate of zinc.—Miller, "Elements of Chemistry," vol. ii. p. 50.

The effect of the copper being coated with hydrogen is that a difference of potential is no longer produced.

Why the hydrogen should appear at the copper, and why it should stop the current, is not well understood.

Amalgamated Zinc.

We have as yet assumed that all the metals used are chemically pure. The ordinary zinc of commerce of which battery plates are made is however not pure, but contains many particles of iron and other metals. If a piece of ordinary zinc be placed in acid, each of these pieces of iron, together with the zinc near it, forms an independent small cell, whose circuit is always closed, whether the main current is closed or open. The currents produced in these small circuits in no way help the main current, while they cause the zinc to be rapidly consumed.

The cost of chemically pure zinc prohibits its use, so a different plan is used, which, though probably first adopted as a make-shift, is found to be in every respect equally efficacious with the employment of pure zinc.

It consists in coating the zinc with mercury. This is done by first immersing the zinc for a few minutes in dilute sulphuric or hydro-chloric acid, so as to give it a chemically clean surface, and then pouring mercury upon it. The mercury at once combines with its surface, and the whole of the zinc appears bright like silver. Zinc thus "amalgamated" is not attacked by dilute sulphuric acid, unless it forms part of a closed galvanic circuit. The precise action of the mercury is not known. It probably acts by coating the zinc and particles of iron alike with one and the same metal.

Binding Screws.

Binding-screws are clamps for attaching connecting wires to any instrument. They are made in many forms, but the two shown in figs. 59, 60 are the most general types. In fig. 59, the end of the wire is passed through the hole, and the screw being turned, clamps it. This form is most convenient for all ordinary apparatus. In the form fig. 60, the wire is bent round and clamped. This is used for measuring apparatus, as it gives rather a better contact. This form is often made with more than one nut, so as to allow a second wire to be attached without disturbing the first.

Fig. 59.

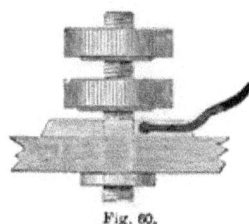

Fig. 60.

CONSTANT BATTERIES.

To make a constant battery or cell, it is necessary to provide some means of freeing the positive or copper plate from hydrogen.

SMEE'S CELL.—FIG. 61.

In Smee's cell, which is shown in section in fig. 61, the plates consist of zinc and platinized silver, i. e. silver with a deposit of rough platinum in powder on its surface. As this presents a multitude of points, the hydrogen disengages itself more easily than from a smooth plate. As silver is much more expensive than zinc, the silver plate is usually arranged between two zinc ones, so as to use both sides of the silver, and so get a greater surface. Although the difference of potential is independent of the size of the plates, we shall explain that the quantity of electricity produced is not.

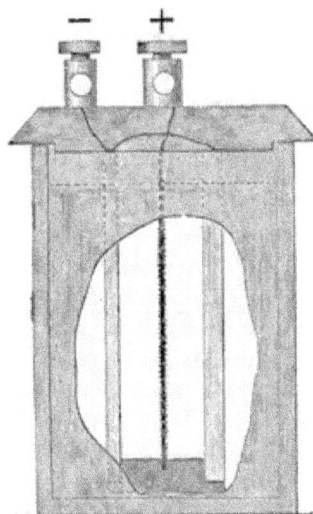

Fig. 61.

THE BICHROMATE OF POTASH CELL.—FIG. 62.

In this cell the plates consist of carbon and zinc, and the

liquid is dilute sulphuric acid saturated* with bichromate of potash. The action of the bichromate of potash is to prevent the hydrogen from reaching the carbon plates at all.†

Owing to its not making any smell, this cell is much used for table work, when a moderately powerful current is required for a short time. It is usually made in what is called the bottle form, shown in section in fig. 62. A globe-shaped glass bottle with a cylindrical neck contains the liquid. The carbon, which, though not called a metal, is an excellent conductor, is what is deposited in the necks of the retorts during the manufacture of coal gas. It is very hard, and can only be ground into shape, not fairly cut. Two plates are made of it, and reach from the bottom to the top of the bottle, where they are fixed to a piece of vulcanite which forms a stopper.

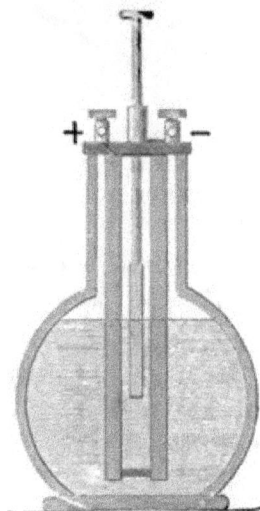

Fig. 62.

A zinc plate only half the length of the carbon plates is fixed to a sliding rod, so that, being still between the carbons, it can be placed either in the bulb or neck of the bottle. The bottle is only filled up to the bottom of the neck, so that when the cell is not in use the zinc can be drawn completely out of the liquid.

On the ebonite stopper are fixed two binding screws which are connected by strips of metal, the one to the sliding rod attached to the zinc, the other to the two carbon plates.

TWO-FLUID CELLS.

We have as yet described only cells with one fluid. In all

---

* About 4 oz. of bichromate will saturate one pint of water. The bichromate should be dissolved in boiling water, and, when it is cold, the acid should be added. When the solution is again cold, it will be ready for use.

† The chemical action is as follows:—

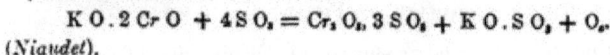

$$K O . 2 Cr O + 4 S O_3 = Cr_2 O_3, 3 S O_3 + K O . S O_3 + O_6.$$

(Niaudet).

these batteries the compounds formed by the hydrogen return to
the zinc plate and retard the action upon it. Cells with two
fluids are made to prevent this taking place. The two principal
types are Grove's and Daniell's cells. The latter is used when a
constant current of moderate strength is required for days, weeks,
or months. The former, when a very powerful current is required
for a few hours.

GROVE'S CELL.—FIG. 63.

In Grove's cell, the metals used are zinc and platinum; the
fluids, strong nitric and dilute sulphuric acids. A cell of thin

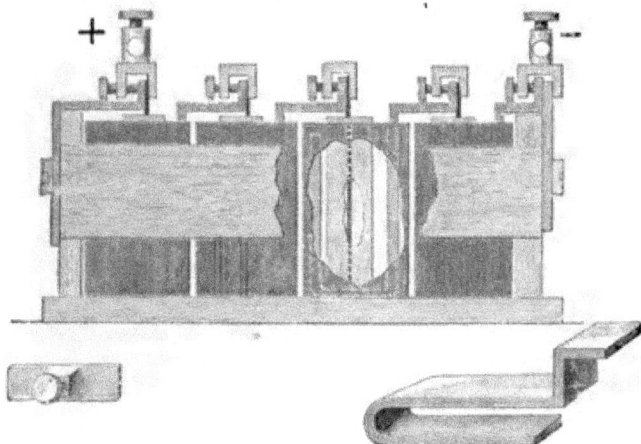

Fig. 63.

porous earthenware is filled with nitric acid, and contains the
platinum plate. This cell is placed inside another cell, usually of
ebonite, containing the zinc and dilute sulphuric acid. The porous
earthenware, when wet, permits the electricity to pass freely
through it, while it almost entirely prevents the liquids from
mixing. In fig. 63, which shows the arrangement of the plates,
several cells are represented connected together, but the reader is
requested for the present to confine his attention to one only. In
this cell, the hydrogen which, wherever it is set free, must be
formed in the sulphuric acid, would have, in order to reach the
platinum plate, to travel through the nitric acid; or even if it is
only liberated on the platinum, it is still in contact with the nitric
acid. The hydrogen and nitric acid at once combine, and form

P

nitrous acid and water, both of which remain in solution in the free acid.*

One of the zinc plates and one of the clamps used for holding the platinum against the zinc (page 221) are shown at the bottom of fig. 63.

Grove's battery is the only voltaic arrangement used for producing the electric light, and for other purposes where great power is required.

BUNSEN'S CELL.

Bunsen's cell is similar in construction to Grove's, with the exception that the plate immersed in the nitric acid is of carbon instead of platinum. The impossibility of cutting carbon into very thin slices necessitates making the cells larger. Usually they are made circular. This cell is not so powerful as Grove's, and, though its first cost is much less, it is troublesome, and more expensive to work with.†

DANIELL'S CELL.—FIGS. 64, 65, 66, 67.

In this cell the metals used are zinc and copper. The former is usually immersed in dilute sulphuric acid; the latter in a saturated solution of sulphate of copper. In a very convenient form of the cell shown in figs. 64, 65, the zinc in the form of a rod is placed inside the porous cell, and the containing vessel being made of copper acts as the other plate.

* The chemical action may be represented by the following formula. The molecular arrangement before the action being represented by the brackets above the line; after it, by those below.

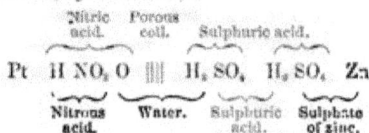

$$\text{Pt} \quad \overbrace{\text{H NO}_3 \text{ O}}^{\text{Nitric acid.}} \mathbin{|\!|\!|} \overbrace{\text{H}_2 \text{SO}_4}^{\text{Porous cell.}} \overbrace{\text{H}_2 \text{SO}_4 \quad \text{Zn}}^{\text{Sulphuric acid.}}$$

Nitric Porous
acid.   cell.   Sulphuric acid.

Nitrous acid.   Water.   Sulphuric acid.   Sulphate of zinc.

We see that the platinum does not join in the chemical action.—Miller, " Elements of Chemistry," vol. i. p. 479.

From his having represented two atoms of sulphuric acid, it appears that Dr. Miller was of opinion that the action between the zinc and sulphuric acid was distinct from that between the sulphuric and nitric acids.

† Both Grove's and Bunsen's cells give off fumes of nitrous acid which are unwholesome, and injurious to instruments, and therefore they must not be placed in the physical laboratory, but in a separate shed or cellar connected to the laboratory by insulated copper wires.

Inside the copper cell and near the top is a copper shelf, perforated with many holes. This shelf serves to keep the porous cell in its place. On it are piled up a number of crystals of sulphate of copper. The cell is filled with a saturated solution of the same, i.e. with water in which is dissolved the maximum quantity of sulphate of copper which it will contain.

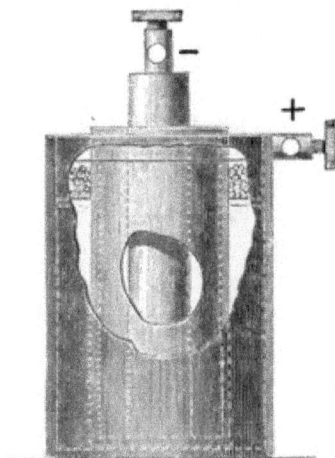

Fig. 64.

Fig. 65.

In the inner cell is the zinc rod, and, according to the purpose for which the battery is required, either dilute sulphuric acid, salt and water, or plain water; the latter, while causing a great diminution of power, increasing the constancy of the battery.

In the cell from which fig. 65 is drawn, the copper cylinder is 6 inches deep and 3 inches in diameter.

When the circuit is closed, the hydrogen, whether it comes from the zinc through the porous cell towards the copper, or is liberated on the copper, meets the solution of sulphate of copper, and, taking from it an atom of sulphur and four atoms of oxygen, forms sulphuric acid, and liberates metallic copper which is deposited on the copper plate. At the same time, sulphate of zinc is formed in the sulphuric acid cell.

Thus the thickness of the zinc plate diminishes, and that of the copper plate increases, while the cell is worked.*

It will be seen, on looking at the equation in the foot-note, that for each molecule of copper deposited on the plate, one molecule of sulphate of copper is destroyed, and the solution gets weaker. As soon as this occurs, a portion of the sulphate of copper on the shelf is dissolved. The part of the liquid which has dissolved it becomes denser than the rest, and sinks to the bottom, and a fresh portion of the weakened solution comes

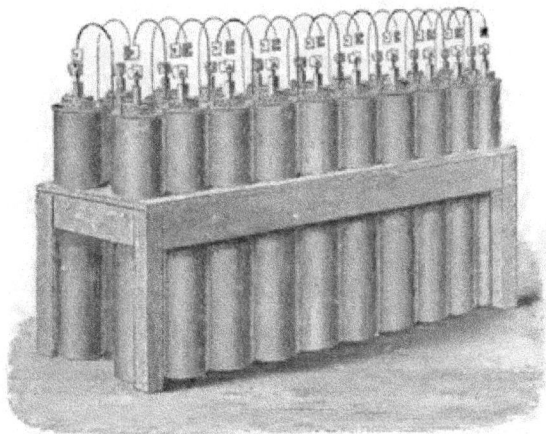

Fig. 66.

in contact with the solid sulphate of copper, and so a circulation goes on till all the liquid is again saturated.

* The following chemical equation represents the change that takes place. The brackets above the line representing the molecular arrangement before the decomposition; those below, the arrangement afterwards:—

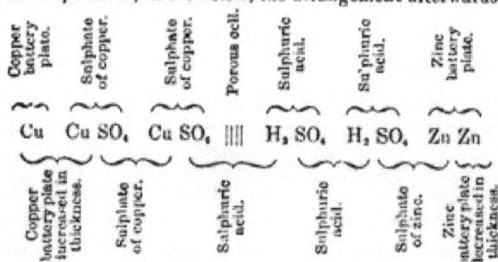

Copper battery plate.	Sulphate of copper.	Sulphate of copper.	Porous cell.	Sulphuric acid.	Sulphuric acid.	Zinc battery plate.
Cu	Cu SO₄	Cu SO₄	‖‖	H₂ SO₄	H₂ SO₄	Zn Zn
Copper battery plate increased in thickness.	Sulphate of copper.	Sulphuric acid.		Sulphuric acid.	Sulphate of zinc.	Zinc battery plate decreased in thickness.

$$Cu \quad Cu\,SO_4 \quad Cu\,SO_4 \quad \|\|\| \quad H_2\,SO_4 \quad H_2\,SO_4 \quad Zn\,Zn$$

Miller's "Elements of Chemistry," vol. i. p. 477.

The power of this cell steadily diminishes until the dilute acid is saturated with sulphate of zinc, after which it remains almost constant for a very long time. For this reason, when constancy is more important than strength, it is customary to saturate the solution with sulphate of zinc before beginning work.

Figs. 66, 67 are drawn from two original models of Daniell's battery, preserved at King's College, where he was Professor of Physics from 1831 to 1845.

The copper cells shown in fig. 66 are 20 inches deep and 3½ inches in diameter.

Fig. 67 shows an arrangement which was adopted for keeping the acid solution constantly renewed, so that its strength should be always constant. A constant supply of fresh acid was allowed to drip in at the top, while the used acid flowed out through the glass tube on the left. This cell is 6 inches deep and 3 inches in diameter. The porous cell is made of parchment. The zinc is not shown. On the right is a mercury cup used instead of a binding screw.

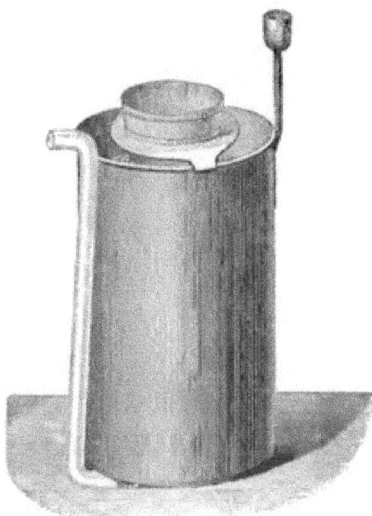

Numerous other forms of Daniell's cell are in use, the various modifications having been introduced with a view of preventing the mixing which goes on through the walls of any porous cell, and because of the resistance which such a cell offers to the electrical and chemical action.

Fig. 67.

GRAVITY BATTERIES—THOMSON'S TRAY CELL, FIGS. 68, 69.

They have usually taken the form of "gravity" batteries, that is batteries where the plates are placed horizontally, and the liquids kept apart, chiefly, if not entirely by their difference in density. The denser liquid is of course placed at the bottom.

The best form, and one which may be taken as a type of the rest,
is Sir Wm. Thomson's tray cell.

The following description of it is taken from Professor Clark
Maxwell's "Electricity," art. 272, vol. i. p. 327 :—

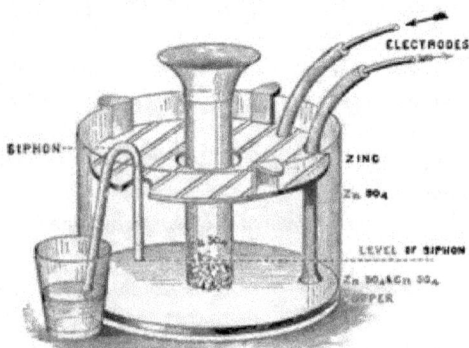

Fig. 68.

"In all forms of Daniell's cell the final result is that the
sulphate of copper finds its way to the zinc, and spoils the battery.
To retard this result indefinitely, Sir W. Thomson[*] has constructed
Daniell's battery in the following form.   In each cell the copper

Fig. 69.

plate is placed horizontally at the bottom, and a saturated solution of sulphate of zinc is poured over it. The zinc is in the
form of a grating, and is placed horizontally near the surface of

* Proc. Roy. Soc., xix., 1870-71, page 253.

the solution. A glass tube is placed vertically in the solution, with its lower end just above the surface of the copper plate. Crystals of sulphate of copper are dropped down this tube, and dissolving in the liquid form a solution of greater density than that of sulphate of zinc alone, so that it cannot get to the zinc except by diffusion. To retard this process of diffusion, a syphon, consisting of a glass tube stuffed with cotton wick, is placed, with one extremity midway between the zinc and copper, and the other in a vessel outside the cell, so that the liquid is very slowly drawn off near the middle of its depth. To supply its place, water, or a weak solution of sulphate of zinc is added above when required. In this way the greater part of the sulphate of copper, rising through the liquid by diffusion, is drawn off by the syphon before it reaches the zinc, and the zinc is surrounded by liquid nearly free from sulphate of copper, and having a very slow downward motion in the cell, which still further retards the upward motion of the sulphate of copper. During the action of the battery, copper is deposited on the copper plate, and SO₄ travels slowly through the liquid to the zinc with which it combines, forming sulphate of zinc. Thus the liquid at the bottom becomes less dense by the deposition of the copper, and the liquid at the top becomes more dense by the addition of the zinc. To prevent this action from changing the order of density of the strata, and so producing instability and visible currents in the vessel, care must be taken to keep the tube well supplied with crystals of sulphate of copper, and to feed the cell above with a solution of sulphate of zinc sufficiently dilute to be lighter than any other stratum of liquid in the cell."

Fig. 68 represents a lecture model, and fig. 69 is drawn from a cell of the form commonly used. The tray in fig. 69 is 22 inches square.

THE LECLANCHÉ CELL.—FIG. 70.

This is now very extensively used for telegraphic purposes. It consists of zinc and carbon separated by a porous cell. The zinc is surrounded by a solution of sal-ammoniac and the carbon by a mixture of black oxide of manganese and powdered carbon. The cell containing the powder is filled up with water. This cell has small power, but for discontinuous work will remain in action, without more attention than occasionally filling up the cells with water, for some years.[*]

Fig. 70.

THE TWO-FLUID BICHROMATE CELL.

In this cell zinc and carbon are used as in the one-fluid form, but the carbon is placed in a porous cell surrounded by a saturated solution of bichromate of potash in water only, while the zinc is placed in weak sulphuric acid (about 20 to 1) in the outer cell. This battery is extensively used for telegraphic purposes.

THE CHLORIDE OF SILVER CELL.—FIGS. 71, 72.

The following is Mr. Warren De La Rue's description of a large battery of these cells :[†]—

"The battery used up till now consists of 1080 cells, each being formed of a glass tube T (fig. 71), 6 inches (15·23 centims.) long and $\frac{3}{4}$ of an inch (1·9 centim.) internal diameter; these are closed with a vulcanized rubber stopper (cork) c, perforated eccentrically to permit the insertion of a zinc rod, carefully amalgamated, $\frac{3}{16}$ (0·48 centim.) of an inch in diameter, and 4·5 inches (11·43 centims.) long. The other element consists of a flattened silver wire sw, passing by the side of the cork to the bottom of the

---

[*] The chemical action is as follows :—

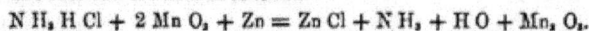

$N H_4 H Cl + 2 Mn O_2 + Zn = Zn Cl + N H_3 + H O + Mn_2 O_3.$

[†] Proc. Roy. Soc., xxiii., 1874-75, p. 356; and Phil. Trans., 1877, vol. clxix. p. 155.

tube, and covered, at the upper part above the chloride of silver
and until it passes the stopper, with thin sheet gutta-percha for
insulation, and to protect it from the action of the sulphur in the
vulcanized corks; these wires are $\frac{1}{16}$ of an inch (0·16 centim.)
broad and 8 inches (20·32 centims.) long. In the bottom of the
tube is placed 225·23 grains (14·59 grms.) chloride of silver in
powder; this constitutes the electrolyte: above the chloride of
silver is poured a solution of common salt containing 25 grammes
chloride of sodium to 1 litre (1752 grains to 1 gallon) of water,
to within about 1 inch (2·54 centims.) of the cork. The con-

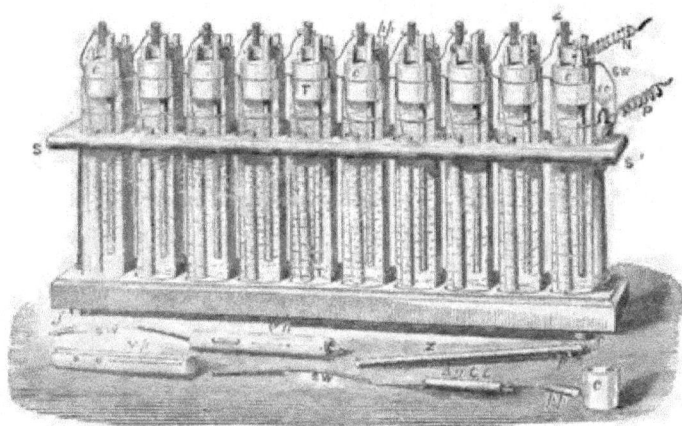

. Fig. 71.

nection between adjoining cells is made by passing a short piece
of india-rubber tube over the zinc rod of one cell, and drawing
the silver wire of the next cell through it so as to press against
the zinc. The silver rod is surrounded by a tube of vegetable
parchment, *rp*, to prevent it touching the zinc. The closing
of the cells by means of a cork prevents the evaporation of water,
and not only avoids this serious inconvenience, but also con-
tributes to the effectiveness of the insulation. More water can
be added through holes in the corks closed by plugs, *pp*.
The tubes are grouped in twenties in a sort of test-tube rack,
S S', having four short ebonite feet, *ff* (fig. 71), and the whole
placed in a cabinet (fig. 72), 2 ft. 7 in. (78·74 centims.) high, 2 ft.
7 in. wide, and 2 ft. 7 in. deep, the top being covered with ebonite

to facilitate working with the apparatus, which is thus placed on it as an insulated table.

The electro-motive force of the battery, as compared with a

Fig. 72.

Daniell's (gravity) battery, was found to be as 1·03 to 1,[*] its internal resistance 70 ohms[†] per cell, and it evolved 0·214 cub. centim. (0·0131 cub. inches) mixed gas per minute when passed through a mixture of 1 volume of sulphuric acid and 8 volumes of water in a voltameter[‡] having a resistance of 11 ohms. The striking-distance[§] of 1080 elements between copper wire terminals, one turned to a point, the other to a flat surface, in air is $\frac{1}{263}$

* See p. 213.   † See Chapter XXVII.
‡ See Chapter XXXI.   § See Chapter XLIV.

inch (0·096 millim.) to $\frac{1}{250}$ inch (0·1 millim.). The greatest distance through which the battery-current would pass continuously in vacuo was 12 inches (30·48 centims.) between the terminals in a carbonic acid residual vacuum. This battery has been working since the early part of November 1874, with, practically, a constant electro-motive force."[*]

POGGENDORF'S CELL.

The following description is taken from M. Niaudet's book on batteries.[†] I have no practical experience of the cell.

The form of the cell is the same as that of Bunsen's. A carbon rod is placed in the porous cell, a zinc cylinder in the outer one.

In the zinc cell is a mixture of 12 parts by *weight* of sulphuric acid to one of water.

In the carbon cell is a mixture (by weight) of

> 100 parts of water,
> 12 ,,   ,, bichromate of potash,
> 25 ,,   ,, sulphuric acid.

This cell is said to have a very large electro-motive force, greater than that of Grove, double that of Daniell.[‡]

BYRNE'S CELL.

Dr. Byrne, of Brooklyn, U.S.A., has invented a cell where the metals are platinum and zinc, and the solution is the same as that in the bichromate cell, p. 207. It is connected to a blow-pipe bellows, and air is forced through the liquid all the time the battery is in operation. This removes the hydrogen. Mr. Ladd, who has described the battery,[§] mentions that the quantity of electricity produced by it is very large.

I have now, I think, described the principal types of batteries in common use. There are, however, an immense number of other cells, differing in various details from those we have described—accounts of them all will be found in books on telegraphy, and in M. Niaudet's work quoted above.

---

[*] Written in 1877.

[†] Niaudet, *Traité Elementaire de la Pile Électrique*, 2nd ed., page 201. (Baudry, Paris, 1880.)

[‡] See table, page 224.

[§] Report, Brit. Assoc., Dublin, 1878, p. 448.

LATIMER CLARK'S STANDARD CELL.

On June 19, 1873, Mr. Latimer Clark communicated to the Royal Society* an account of a " Standard Cell," that is, a cell whose electro-motive force is always constant. Great difficulty had been experienced in determining a practical unit of electro-motive force, owing to the fact that not only are there differences in the electro-motive forces of different ordinary cells, supposed to be of the same construction, but that the electro-motive force of the same cell varies from day to day.

With the " standard cell " it is found that as long as it is not used to produce a current, the difference of potential between its poles remains absolutely constant. The maximum difference observed in a series of comparisons between different models of it during a period of several months was not more than $\frac{1}{1000}$ part of the whole electro-motive force, and it appears that even this difference might be accounted for by an accidental difference of temperature.

TEMPERATURE.

The electro-motive force which the cell gives at 15°·5 C. is taken as the standard.†

It is found that the force decreases with increase of temperature, and that the rate of variation for 10° above and below 15°·5 is 0·6 per cent. for each degree centigrade.

CONSTRUCTION.

" The battery is formed by employing pure mercury as the negative element, the mercury being covered by a paste made by boiling mercurous sulphate in a thoroughly saturated solution of zinc sulphate, the positive element consisting of pure distilled zinc resting on the paste.

" The best method of forming this element is to dissolve pure zinc sulphate to saturation in boiling distilled water. When cool, the solution is poured off from the crystals and mixed to a thick paste with pure mercurous sulphate, which is again boiled to drive off any air; this paste is then poured on to the surface of the mercury, previously heated in a suitable glass cell; a piece of pure zinc is then suspended in the paste, and the vessel may be advantageously sealed up with melted paraffin wax. Contact

* Phil. Trans., 1874, page 1.
† It equals 1·457 Volt. See Chapter XXI.

with the mercury may be made by means of a platinum wire
passing down a glass tube, cemented to the inside of the cell,
and dipping below the surface of the mercury, or more conveniently by a small external glass tube blown on to the cell,
and opening into it close to the bottom. The mercurous sulphate (Hg_2SO_4) can be obtained commercially;[*] but it may be
prepared by dissolving pure mercury in excess in hot sulphuric
acid at a temperature below the boiling-point: the salt, which
is a nearly insoluble white powder, should be well washed in distilled water, and care should be taken to obtain it free from the
mercuric sulphate (persulphate), the presence of which may be
known by the mixture turning yellowish on the addition of
water. The careful washing of the salt is a matter of essential
importance, as the presence of any free acid, or of persulphate, produces a considerable change in the electro-motive
force of the cell."

BATTERIES OF SEVERAL CELLS.

We have said that when the circuit is open (that is, when the
poles are not connected), the potentials of the poles of any cell differ
by a quantity which is approximately constant for each kind of cell.
We often, however, require a difference of potential greater than
can be given by any one cell. This is obtained by connecting a
number of similar cells " in series," that is, connecting the positive pole of one with the negative pole of the next, and so on—
a number of *cells* so connected is called a voltaic *battery*. Fig. 63
is a representation of a Grove's battery of four cells. It is seen
that the zinc of each cell projects sideways over the next, and
the platinum of that cell is clamped to it. The only reason
why the zinc plates are chosen to project rather than the platinum, is the far greater expense of the latter, and the fact that
owing to their not being consumed, it is only necessary to make
them of the thickness of writing-paper, when of course they
have but small rigidity.

Thus all the poles are connected two and two, except one from
each of the end cells. These two free poles are called " the poles
of the battery."

Their difference of potential is as many times the difference of

. [*] Mr. Clark has obtained it from Messrs. Hopkins and Williams, 5, New
Cavendish Street.

potential between the poles **of a** single cell, as **there are cells in**
the battery, i. e., in a battery of 4 cells, if we suppose the difference of potential between two poles of the same cell to be represented **by the number** 10, that between the poles of the battery
will be represented **by 40; if there are** 5 cells, **by 50, and**
so on.

For let us suppose that **the negative pole of** the end cell (on
the right hand **in fig. 63) is connected to the** earth; its
potential is zero. The potential of the positive pole will then be
10. But the positive pole of the first cell is in metallic communication with the negative pole of the second, and so their
potentials are equal,* and therefore the potential of the negative
pole of the **second cell is 10.** But the common difference of
potential being 10, the positive pole of **the second cell has a**
potential of 20. This is in metallic communication with the
negative pole of the third cell, **and** therefore the potential **of the**
positive **pole of** that cell is 30, and that of the positive pole **of**
the fourth cell 40, or the difference between the potentials of the
poles of the 4-cell battery is 40, or four times the difference
between the poles of each cell.†

---

* If we consider the difference of potential to take place at the contact of
the metals, we must consider the potentials of the two metals in the same
cell to be equal, **and the above argument will** still hold.

† ON THE CARE OF A GROVE BATTERY.

When a Grove's battery is **much used, it** is worth while **to** make special
arrangements for keeping it in good order, and to facilitate setting it up and
taking it to pieces. The following plan has been adopted with success by the
present writer :—

The zincs are kept in a large earthenware "crock" full of water, in which
a large quantity of common soda has been dissolved. The clamps are kept in
a jam-pot full **of** the same. The porous cells are kept in a large crock full of
pure water.

On commencing to set up the battery, **as many porous cells as** are wanted
are taken out of the water and set wrong side up to **drain.** The ebonite cells
are **then** about half-filled with dilute sulphuric acid. **The** zincs are taken out
of the soda, and the face against which the platinum is to be pressed is dipped
(without rinsing) for a few moments in dilute acid, and then into mercury,
and rubbed with an old tooth-brush.

This produces a clean metallic surface. **The** surfaces of the two *end* clamps
are treated in the same way. The zincs are put in to the ebonite cells, and
the porous cells filled with strong nitric acid and placed inside the zincs. The
platinums are put in and clamped as in fig. 63, and the ebonite cells filled up
with dilute sulphuric **acid.** The battery is now ready for work. It must

EFFECTS OF THE ELECTRIC CURRENT.

HEATING EFFECTS.

The electric current heats the wires in which it passes. The amount of heating depends on the length, thickness, and nature of the wire, and on the strength of the current. See Chapter XXXIV.

CHEMICAL EFFECTS.

The electric current, when passed through certain solutions of compound chemical substances, decomposes them into their component elements. These actions will be treated of under the heading Electrolysis. See Chapter XXXI.

MAGNETIC EFFECTS.

The magnetic effects of the current will be treated of in Chapter XX.

CONVENTIONAL SIGN.

To save repeating pictures of the battery, the conventional sign (fig. 73) is used in the diagrams; the thin and thick lines representing respectively the zinc and other plates, and the number of them the number of cells.

Fig. 73.

not be placed in the physical laboratory, but in a separate battery-room, as the fumes are both poisonous, and injurious to instruments.

In taking the battery to pieces, the clamps are thrown first into their jar. Then the platinums are rinsed under a tap, and placed, without wiping, in their box. Then the porous cells are emptied into a bottle provided with a large funnel and put into the water jar, and then the zincs put into the soda. The sulphuric acid is left in the ebonite cells.

As all the operations of taking down a battery can be done with one hand, much time may be saved by working at two cells simultaneously with the two hands. A skilful operator should be able to set up a 10-cell Grove's battery in about six minutes, and to take it down in about a minute and a half.

There is no economy in purchasing small cells, as with them the acids can seldom be used more than once, whereas, with large ones, they can be used four or five times. "Quart cells," with platinums 6 inches by 3 inches, are the best size.

New zincs will require amalgamating every day for 4 or 5 days; after that about once a fortnight till they are worn out. If possible the instrument maker should be persuaded not to put any paint on them.

ELECTRO-MOTIVE FORCES OF **VARIOUS CELLS.**

The following **table** is **given by** M. Niaudet,[*] and compares the electro-motive forces of **all** the different cells described in his book. The volt is the common unit of electro-motive force (see vol. i., page 260). I have reduced the measures of sulphuric acid from parts by weight to parts by volume.

ELECTRO-MOTIVE FORCES.

					Volts.
Daniell	Zinc amalg.	Sulphuric acid, 7½ to 1	Saturated solution of copper sulphate	Copper	1·079
,,	,,	22 to 1	,,	,,	0·978
,,	,,	,,	Nitrate of copper saturated	,,	1·000
,,	,,	,,	Sulphate of copper	,,	0·909
,,	,,	Sulphate of zinc	,,	,,	0·955
,,	,,	1 part common salt, 4 parts water	,,	,,	1·060
Grove	,,	Sulphuric acid, 7½ to 1	Nitric acid (fuming)	Platinum	1·956
,,	,,	Salt water	Nitric acid, sp. gr. 1·33	,,	1·904
,,	,,	Sulphuric acid, 22 to 1	,,	,,	1·810
,,	,,	Sulphate of zinc	,,	,,	1·672
Bunsen	,,	Dilute sulphuric acid	Nitric acid	Carbon	1·734
Callan	,,	,,	,,	Cast iron	1·700
Poggendorf	,,	,,	Chrome mixture	Carbon	{1·796 {2·028
Marié Davy	,,	Sulphuric acid, 22 to 1	Paste of sulphate of mercury	,,	1·524
,,	,,	Dilute sulphuric acid	,,	,,	1·33
Leclanché	,,	Solution of sal ammoniac	Binoxide of manganese	,,	1·481
De La Rue	Zinc	Chloride of silver		Silver	1·029
Becquerel	Zinc amalg.	Sulphate of zinc	Sulphate of lead	Lead	0·55
Niaudet	,,	Common salt	Chloride of lime	Carbon	1·65
Duchemin	,,	,,	Perchloride of iron	Lead	1·541
,,	Platinum	Dilute sulphuric acid	Dilute sulphuric acid	Platinum	1·79
Planté	Lead	,,	,,	Lead	2·5
Latimer Clark, Standard cell	Zinc amalg.	Sulphate of zinc	Paste of sulphate of mercury	Mercury	1·457

Porous Cell.

* *Pile Électrique*, 2nd ed., page 256.

CHAPTER XIX.

ELECTRICITY OF CONTACT.

WHEN two different metals are in contact, there is, in general, an electro-motive force acting across the junction from the one to the other.

For instance, if a piece of copper and a piece of zinc be soldered together, the zinc will be found to be positive as compared with the copper. Volta's theory of contact electricity is based on this fact.

This electro-motive force cannot in general produce a current, for to form a closed circuit of the two metals, two junctions are necessary, and the electro-motive forces at the two junctions will be equal and opposite. It is found that the insertion of an intermediate metal does not destroy the balance, for the following law holds:—

Let A B C be any three metals* arranged in circuit

$$c\overset{A}{\bigvee}{}_{\!B}$$

then the junctions are A B, B C, C A, and the electro-motive forces at them may be written

$$F_{A B}, F_{B C}, F_{C A};$$

then always we shall have, if A, B, and C are at the same temperature,

$$F_{A B} + F_{B C} + F_{C A} = 0;$$

or, the electro-motive force between A and B is equal and opposite to the sum of the electro-motive forces between B and C and C and A, and the same is true for any number of metals in circuit. It is obvious that if this were not so, the law of the conservation of energy would not hold; for we could obtain a current without chemical or mechanical action.

The neglect of this limitation led Volta and his followers into

---

* C may be the wire of a galvanometer.

such absurdities, that Faraday was induced to deny the existence of contact electricity altogether.

A large portion of the second volume of his "Experimental Researches" is devoted to proving that all cases of supposed electrification by contact can be shown to be due to either chemical or mechanical action.

Faraday's mistake is easily explained. The electrometer of his day (1839) being a very untrustworthy instrument, he used a galvanometer, which would not show the existence of a simple difference of potential, but only that constant renewing of the difference which we call current. In all these cases he rightly said that the effects observed could always be traced to some chemical or mechanical cause. It is now well known that, *though contact produces a difference of potential, this difference of potential only produces a current when some extraneous means are employed to keep it constantly renewed.*

In the Voltaic battery, according to Volta's theory, the action of the liquid is to reduce the two metals to the same potential.

The difference of potential being at once renewed at the junction, a continuous current is kept up at the expense of the chemical action between the liquid and one of the metals.

Professor Maxwell quotes this theory* without expressing any opinion about it. Sir Wm. Thomson,† however, says, " For nearly two years I have felt sure that the proper explanation of Voltaic action in the common Voltaic arrangement is very near Volta's . . . . I now think it is quite certain that two metals, dipped in one electrolytic liquid, will (when polarization is done away with) reduce two dry pieces of the same metals, when connected each to each by metallic arcs, to the same potential."

Instead of equalizing the potentials chemically, we may do it mechanically, as has been done by Sir Wm. Thomson.‡

A copper funnel is fixed into an insulated zinc tube, as shown in fig. 74. The contact between the copper and zinc produces a difference of potential. Copper filings placed in the funnel

* "Electricity," **247, vol.** i. p. 300.

† "Proceedings **Lit. and** Phil. Soc. of Manchester," Jan. 21, 1862, and "Papers on Electro-statics," p. **318.**

‡ Proc. Roy. Soc., 1867, vol. xvi. p. 71, and "Papers on Electro-statics," p. 324.

acquire the same potential as the funnel, and therefore their potential differs from that of the zinc tube. They are allowed to stream out through the tube without touching it. Each as it falls becomes negatively electrified by induction, and they produce a rapidly increasing negative charge on a small insulated can, placed to catch them. Now, if the can be connected to earth by a wire, a current will flow through that wire as long as the stream of filings continues. In this case the difference of potential is undoubtedly caused · by contact, but the energy required to convert this difference of potential into current is supplied by the work done by gravity on the falling filings.

Fig. 74.

On June 15th, 1876, Messrs. Hugo Müller* and Warren De La Rue communicated to the Royal Society an account of an apparatus where the energy required to enable the electrification of contact to produce a current was obtained by a piece of mechanism which " brings together and separates two discs, one of copper and one of zinc, each six inches diameter, 400 times in a minute, and after each separation makes the zinc plate touch a spring attached to an insulated conductor ; and, moreover, by means of cams, makes earth connection with either disc, or with both, previous to their being brought again into contact."

They found that when the apparatus was making 320 breaks a minute, the tension of the electricity as compared to that of a chloride of silver cell was as

30·88 to 1,

that is, that when the machine was connected to the electrometer, the deflection was nearly equal to that which would have been produced by 31 cells in series.

A feeble current was obtained when the electricity was led to earth through a reflecting galvanometer ; it gave 35 divisions of the scale, or about $\frac{1}{140}$ part of that produced by ½-inch bits of copper and zinc wire, held one in the right hand and one in the left between dry fingers.

Mr. Joseph Thomson† states that he has found that if cakes

---

* Proc. Roy. Soc., 1876, vol. xxv. p. 258.
† Ibid., June 15, 1876, vol. xxv. p. 169.

be made, each of two insulating substances, and the electrified needle of an electrometer be suspended over them along their line of separation, it will be deflected, showing a difference of potential between them.

He has found that, in the following cakes, the substance first mentioned becomes (+), the second (−).

(+)		(−)
Glass	and	wax,
Glass	„	resin,
Glass	„	sulphur,
Glass	„	solid paraffin,
Zinc	„	sulphur,
Sulphur	„	ebonite.

He observes :—

" The series so far, being in the same order as the frictional series, seems to suggest that the electrical displacement which takes place when two non-conductors are put in contact, acts as a predisposing cause, in virtue of which the work done in rubbing them together is converted into electrical separation."

Numerous other experiments have been made, but without decided results up to 1876.

EXPERIMENTS OF AYRTON AND PERRY.—PLATE XX.

At the meeting of the British Association in 1876, Professors Ayrton and Perry communicated a preliminary notice of a series of experiments they had made to determine whether in a galvanic cell—for example, a Daniell's, Grove's, &c.—the electro-motive force of the cell was, or was not, equal to the algebraical sum of all the differences of potential, each being measured separately, at the various contacts of dissimilar substances in the cell. A further account of the investigation appeared subsequently in Parts I. and II. of the " Contact Theory of Voltaic Action,"* and the experiments were fully described, by which the following law was proved :—

" The electro-motive force of contact of two metals or two electrolytes, or of a metal and an electrolyte, is in each case a constant for the same temperature and in the same gas ; that is to say, if AB means the electro-motive force of contact of the metal or electrolyte A, and the metal or electrolyte B (measured when A and B are not in contact with other conducting substances), AB being identical with—BA ; then the total electro-motive force of any

* Proc. Roy. Soc., vol. xxvii., 1878, p. 196.

closed heterogeneous circuit composed of the substances A, B, C, N is:

$$AB + BC + \&c. + MN + NA."$$

They go on to say :—"The proof of this law was very important, as it was generally thought not to hold true.".

"For example, Professor F. Jenkin, in the edition of his ' Electricity and Magnetism' of 1873, said, on page 44 :— ' The following series of phenomena occur when metals and an electrolyte are placed in contact:—1. When a single metal is placed in contact with an electrolyte, a definite difference of potentials is produced between the liquid and the metal. If zinc be plunged in water, the zinc becomes negative, the water positive. Copper plunged in water also becomes negative, but much less so than zinc. 2. If two metals be plunged in water (as copper and zinc), the copper, the zinc, and the water forming a galvanic cell, all remain at one potential, and no charge of electricity is observed on any part of the system."

If *all* the substances A, B, C, &c., N are metals, then

$$AB + BC + \&c. + MN = AN;$$

but if one or more of them be electrolytes, solid or liquid, then Prof. Ayrton and Perry's experiments show that the difference of potential between A and N when joined by B, C, &c., is equal to

$$AB + BC + \&c. + MN;$$

but we cannot from this sum predict the value of AN, the difference of potentials between A and N when joined *directly*, since the difference of potentials *depends on the path taken* in going from one body to another when electrolytes are in question.

In Messrs. Ayrton and Perry's experiments an *Induction* method was used which got rid of all the difficulties caused by the *contact* of the substances under examination with the poles of the electrometer.

The method of measurement was as follows :—Let 3 and 4 (fig. 75) be two insulated gilt-brass plates connected with the electrodes of a delicate quadrant electrometer. Let 1 under 3 and 2 under 4 be the surfaces whose contact difference of potential is to be measured. 3 and 4 are first connected together and then insulated, but remain connected with their respective electrometer quad-

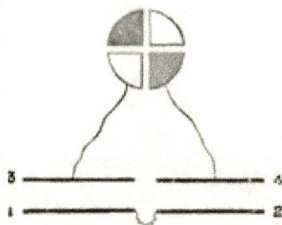

Fig. 75.

rants. Now, 1 and 2 are made to change places with one another,

1 being now under 4 and 2 under 3, then the deflection **of the electrometer needle** will give **a measure** of the difference **of potentials** between 1 and 2. And **it is shown** theoretically that "the difference of potentials d observed with the electrometer will be proportional to the difference of potentials a that we desire to measure, provided the induction arrangement is symmetrical in both positions, or provided that A, B, and D be each nought, where A and A $+$ a **are** the differences of potential of 1 and 2, B the common potential of **3 and 4 before** reversal, and D and D $+$ d their respective potentials **after reversal.** Perfect symmetry in the apparatus being impossible, the latter condition was always experimentally fulfilled."

The actual apparatus used **in** 1876 somewhat improved for the subsequent investigation, and described **by** the authors in their **paper** No. III. of 1879,* is shown in Plate XX.

The substances of which the contact difference of potential are to be measured are **carried on a** table, AB.

In Plate **XX. a** liquid, **L, and** a solid plate, P, of about 530 **sq. centims.** area are shown in position. The table A B is levelled **by three** screws, ll, **and carried** on three brass wheels, W, which **run on a** circular **very rigid** metallic railway, R. To avoid lateral **motion the table is kept** centred by a stout iron pin M, **turning in a brass socket S.**

In order to rotate the lower substances, 1 and **2, it is** necessary, if one or both of them be a liquid, **first** to increase the distance between **1, 2** and **3, 4 in order that 3 and** 4 **may** not strike against the sides of the vessel carrying the liquid.

This was done by raising and lowering the upper plates, 3, 4, by means of the "parallel ruler motion" shown in Plate XX.

The upper framework is lifted by the rod rr, which has a cross-piece pp, which can either be lowered through the slot ss, or by turning the rod rr caused to rest across it.

The upper plates **are supported** by **clean glass rods G, which are kept** dry by sulphuric **acid in the** lead cups U.

The whole apparatus, including **the** short circuit key and elec-**trometer, was,** to avoid induction from outside, enclosed in a large zinc **case connected with the earth, and** was not opened at all during one **complete experiment, consisting** of some ten short circuitings **of the upper plates, reversals** of the table A B, and corresponding **readings to the right and left of the** electrometer needle.

* Phil. Trans., 1880, p. 15.

The following is a complete operation to obtain the contact difference of potentials, between a metal and liquid, for example. Suppose the permanent adjustments to have been made, and the gilt plates 3 and 4 are quite bright. The plate P is cleaned with emery paper that has touched *no other metal*, and all traces of the emery removed by means of a clean dry cloth; it is then placed on the three levelling screws *l*, and fixed in position by hole, slot, and plane.* The porcelain dish containing the liquid is laid in a metal one just fitting it, and on the base of which is a hole, slot, and plane; this is now laid on the other levelling screws *l*.

The rod *r r* is then lowered until the disk *d d* rests on a brass plate let into the top of the wooden framework at the top of the instrument—that is, until the induction plates 3 and 4 are in their lowest position. The levelling screws *l l* are now raised until a small metal ball, of a diameter of eight millims., is in contact at three fixed points with the plate 4 and the plate P, or until, when in contact with the plate 3, it and its reflection in the liquid L appear to meet. To avoid any harm arising from possible contact of the liquid with this gauge ball, it was made of a material not acted on by the particular liquid under experiment.

Before proceeding further, each pair of quadrants is in succession put to earth, the other pair remaining insulated in order to test for any possible leakage from the needles to the quadrants. Each pair of quadrants is now charged with a battery, the other pair being connected with the earth, in order to test for any leakage along the glass rods G, the small glass rods supporting the quadrants in the electrometer, or along the paraffined ebonite pillars of the short circuiting key. It having thus been ascertained that there is no leakage, the strip of metal which has been cut from the same sheet of metal as P itself, and temporarily attached to it by a binding screw soldered to P, is made quite bright with emery paper and a cloth, and its end is dipped into the liquid L, as shown in Plate XX., fig. 4. The zinc case is then closed up, plates 3 and 4 connected together, and with the earth, by means of a key (the handle of which was a long

---

* "Hole, slot, and plane." This is an arrangement invented by Sir Wm. Thomson to allow any apparatus supported on three feet to be removed from a table and replaced exactly in the same position. Let 1, 2, 3 be the feet; 1 is placed in a small *hole* made in the table; 2 in a short *slot* whose direction if produced would pass through the hole; 3 rests on the *plane* surface of the table.

thin ebonite rod projecting through the zinc case), and the electrometer reading taken. 3 and 4 are then insulated from one another, and from the earth, and raised by means of the rod $r\,r$ projecting above through the zinc case ; the table A B is turned from below by means of a handle passing through the base of the instrument ; 3 and 4 are then lowered into exactly their former position, this being ensured by the parallel motion of the supporting beam and by the limiting stop, $d\,d$. The reading of the electrometer is now taken. Then the processes of " short circuit, insulate, raise, reverse, and lower, and take a new electrometer reading, &c.," are repeated.

Some ten readings having thus been obtained, a fresh set of experiments is always made with the same two substances in the following way in order to compensate for the error introduced by defects in parallelism of the apparatus affecting the result obtained from two rigid surfaces (as those of copper and zinc), differently from the result found with one or with two liquid surfaces under examination. Instead of commencing, as before, with the liquid L under 3 and the plate P under 4, the experimenters start with the plate under 3 and the liquid under 4, and readjust, by means of the levelling screws l, the heights of the surfaces, until their distance from the plates 3 and 4 is, as before, 8 millims. They then short circuit, insulate, raise, reverse, and lower, and take exactly as many readings as before ; and the mean of the two sets of readings, obtained with the two modes of levelling, is regarded as the result of the particular experiment.

" To test the accuracy of the statement, quoted on page 229 from Professor Jenkin's 'Electricity,' that when copper and zinc are both plunged into water they are all at the same potential, the following sets of experiments were made. The plates 1 and 2 (fig. 75) were respectively zinc and copper, and they were connected together by means of a liquid in a small beaker having no direct inductive action on the plates 3 and 4. First, however, the apparatus was calibrated thus :—

" TABLE VIII.—13th April, 1876. Plates 10 mm. apart.
Latimer Clark's Standard Cell.

Zero.	Reading.	Deflection.
955·0	892·0	63·0
954·5	1018·5	64·0
954·5	891·8	62·7
953·1	1017·1	64·0

Mean . . . 63·4
Assumed to be 1·457 volts.
Direct reading is 355
Therefore ratio is $\frac{355}{63·4}$ or 5·6.

Experiments :—
" Zinc and copper connected by distilled water at 17° C. Zinc is
negative to copper.

Zero.	Reading.	Deflection.
953	960·2	7·2
952	947·0	5·0
952	960·0	8·0
952	946·5	5·5
951·9	961·0	9·1
952	945·0	7·0
952	961·0	9·0
952·9	946·2	6·7

An interval of 15 minutes.

953	961·0	8·0
952·8	945·1	7.7

Mean . . 7·32 or 0·168 volts.

" Zinc and copper metallically connected. Zinc positive to copper.

Zero.	Reading.	Deflection.
953·0	926.0	27·0
952·7	990·0	37·3
951·0	920·3	30·7
950·1	985·1	35·0
950·0	919·5	30·5
950·2	984·6	34·4
951·0	918·0	33·0
951·1	985·2	34·1

Mean . . 32·7 or 0·751 volts.

" Zinc and copper connected by saturated pure zinc sulphate at
17° C. Zinc negative to copper.

Zero.	Reading.	Deflection.
952·0	961·5	9·5
951·9	944·2	7·7
951·8	960·0	8·2
951·9	943·1	8·8
952·0	960·0	8·0
952·1	944·6	7·5

Interval of 10 minutes.

953·1	960·0	6·9
953·2	945·0	8·2
953·1	961·2	8·1
953·3	945·2	8·1
953·7	961·0	7·3
953·9	945·2	8·7

Mean of first six . . 8·3 or 0·191 volts.
Mean of last six . . 7·9 or 0·182 volts.

" From these experiments it followed that the above statement made in text-books, and which was based on certain experiments of Sir William Thomson, is only approximately correct."

From Professor Ayrton and Perry's experiments, and from those previously made by Sir William Thomson, they were led to conclude " that, when zinc and copper are immersed in water, there are three successive states to be noticed. At the instant of immersion the zinc and copper may possibly be reduced to the same potential, so that the electro-motive force of the voltaic cell E is equal to the difference of potential ZC between zinc and copper in contact ; the zinc now becomes negative to the copper, so that E reaches a limit which is greater than ZC ; lastly, if a current be allowed to pass by metallically connecting the zinc and copper, polarization occurs and the zinc becomes gradually less negative to the copper, E diminishing, therefore, from its maximum value. But when a saturated solution of zinc sulphate is employed instead of water, the first state, if it exists at all, exists for so short a time that practically zinc and copper in zinc sulphate are never at the same potential. Thus when care is taken to keep the zinc and copper in a water cell well insulated from one another, E is found to increase from a value very little greater than ZC, the electro-motive force of contact of zinc and copper, to a limit, but in a zinc sulphate cell no such great increase is observed."

Subsequently the difference of potentials of a number of single contacts of dissimilar substances were measured, as well as the electro-motive forces of complete and incomplete cells built up with the *very same specimens* of the materials immediately after the previous tests were made. The following are some of the results obtained :—Let C, Z, and L represent the copper, zinc, and liquid respectively of a simple cell let L_1 and L_2 be the liquid in contact with the copper, and the liquid in contact with the zinc of a Daniell's cell; let CL be the electro-motive force of contact of C and L, and let CL be identical with—LC. Then :—

I. Daniell with pure saturated copper sulphate and nearly pure saturated zinc sulphate.

Observed EMF of cell
measured directly.

$$CL_1 + L_1L_2 + L_2Z + ZC$$
$$= 0·028 - 0·033 + 0·358 + 0·750 = 1·103$$

1·068 to 1·081, increasing slowly.

II. Daniell with distilled water and pure saturated copper sulphate.

$$CL_1 + L'L_2 + L_2Z + ZC$$
$$= 0·028 + 0·071 + 0·126 + 0·750 = 0·975$$

0·995.

III. Daniell with very dilute zinc sulphate and slightly impure saturated copper sulphate.

$$CL_1 + L_1L_2 + L_2Z + ZC$$
$$= 0 + 0·063 + 0·177 + 0·750 = 0.990$$

1·010.

IV. Simple cell, nearly pure saturated zinc sulphate.

$$CL + LZ + ZC$$
$$= -0·113 + 0·358 + 0·750 = 0·995$$

1·000.

V. Simple cell, distilled water.

$$CL + LZ + ZC$$
$$= 0·074 + 0·126 + 0·750 = 0·950$$

0·832 to 0·942 increasing slowly.

" In every case the sum of the separate contact electro-motive forces is so nearly equal to the observed maximum electro-motive force of the cell, that we have good reason for concluding that the electro-motive force of contact of any two substances measured inductively is constant for *exactly* the *same* specimens of the materials under *exactly the same condition* as regards temperature, the gaseous medium surrounding them, &c., and is quite independent of any other substances that may be in the circuit."

In the investigation made during 1877–78, and described in their third paper,[*] the authors have obtained the following results :—

* Phil. Trans. Roy. Soc., 1880, p. 15.

MEAN CONTACT DIFFERENCES OF POTENTIAL IN VOLTS; SOLIDS WITH SOLIDS IN AIR.†

	Carbon.	Copper.	Iron.	Lead.	Platinum.	Tin.	Zinc.	Amalgamated zinc.	Brass.	
Carbon . . .	0	·370	·485*	·858	·113	·795*	1·096	1·208*	·414*	
Copper . . .	—·370	0	·146	·542	·238	·456	·750	·804	·087	
Iron . . .	—·485*	—·146	0	·401*	·369	·313*	·600*	·744*	—·064	The average temperature at these experiments were made was about 18° C.
Lead . . .	—·858	—·542	—·401*	0	·771	·099	·210	·357*	·472	
Platinum . .	—·113*	—·238	—·369	·771	0	·690	·981	1·125*	·287	
Tin . . .	—·795*	—·456	—·313*	·099	—·690	0	·281	·463	·372	
Zinc . . .	—1·096	—·750	—·600*	·210	—·981	—·281	0	·144	—·679	
Amalgamated zinc .	—1·208*	—·804	—·744	—·357*	—1·125*	—·463	—·144	0	—·822*	
Brass . . .	·414	—·087	·064	·472	—·287	—·372	·679	·822*	0	

The numbers without an asterisk were obtained directly by experiment, those with an asterisk by calculation; using the well-known assumption that in a compound circuit of metals all at the same temperature there is no electro-motive force.

The numbers in a vertical column below the name of a substance are the differences of potential,in volts, between that substance and the substance in the same horizontal row as the number, the two substances being in contact. Thus lead is positive to copper, the electro-motive force of contact being 0·542 volt.

The metals were those of commerce, and therefore not chemically pure.

The authors point out that the contact difference between copper and zinc, which they find to be exactly ·750 volt, is a more convenient and reliable standard of electro-motive force, when it can be used, than even a Latimer Clark's cell.

† In the paper it is explained that in all probability if these quantitative experiments were made in other gases than air very different results would be obtained.

MEAN CONTACT DIFFERENCES OF POTENTIAL IN VOLTS; SOLIDS WITH LIQUIDS, AND LIQUIDS WITH LIQUIDS (IN AIR).

Solutions.	Carbon.	Copper.	Iron.	Lead.	Platinum.	Tin.	Zinc.	Amalgamated zinc.	Brass.	Mercury.	Distilled water.	Alum solution, saturated at 16° C.	Copper sulphate solution, saturated at 15°.3 C.	Zinc sulphate solution, saturated at 15°.3 C.	Zinc sulphate solution: specific gravity 1·125 at 16°·9 C.	1 distilled water, 3 saturated solution zinc sulphate.	Strong nitric acid.
Mercury	·092 {·91 to ·17*}	·808 {·289 to 1·06}	·502	{...} ·171	1·56 {·234 to 2·45}	·177	−·105 to +·156	·100	·251		·043			·164			
Distilled water		−·127	−·118	−·189	·216	−·225	−·536		−·014				·043	·066			
Alum, saturated at 16°·5 C.		·050	−·653														
Copper sulphate, saturated at 16° C.		−·103													·066		
Copper sulphate solution; specific gravity 1·087 at 16°·6 C.																	
Sea salt, specific gravity 1·18 at 26°·5 C.		−·475	−·005	−·267	·856	−·334	−·565		−·435		−·161		−·065				
Sal ammoniac, saturated at 15°·5 C.		·396	−·652	−·189	·067	−·364	−·637		−·348								
Zinc sulphate, saturated at 15°·3 C.							−·499										
Zinc sulphate solution; specific gravity 1·125 at 16°·9 C.							−·239	−·284					−·093				
1 distilled water, mixed with 1 saturated solution zinc sulphate.																	
1 distilled water, mixed with 3 saturated solution zinc sulphate.							−·444						−·102			·102	

* Depending on the carbon.

MEAN CONTACT DIFFERENCES OF POTENTIAL IN VOLTS; SOLIDS WITH LIQUIDS, AND LIQUIDS WITH LIQUIDS (IN AIR)—*continued*

	Carbon.	Copper.	Iron.	Lead.	Platinum.	Tin.	Zinc.	Amalgamated zinc.	Brass.	Mercury.	Distilled water.	Alum solution, saturated at 15° C.	Copper sulphate solution, saturated at 15° C.	Zinc sulphate solution, saturated at 15°·3 C.	Zinc sulphate solution; specific gravity, 1·135 at 16°·9 C.	1 distilled water, 5 saturated solution zinc sulphate.	Strong nitric acid.
Dilute sulphuric acid — Distilled water, with a slight trace of sulphuric acid.	About -·93						-2·41										·93
Distilled water, with about one-fifth per cent. of sulphuric acid.							-2·39										
Distilled water, with about one-third per cent. of sulphuric acid.							-3·12										
By weight— 20 distilled water, 1 strong sulphuric acid	·01 to ·3						-3·44	-3·28									
By volume— 10 distilled water, 1 strong sulphuric acid								-4·29									
By weight— 5 distilled water, 1 strong sulphuric acid	·53 to ·85 *			-·129	1·600 to 1·900	-·256		3·48	·916		1·295	1·436	1·299	1·699			
1 distilled water, 5 strong sulphuric acid		1·413		730 to 1·282	·672					·475							
Concentrated — Sulphuric acid																	
Nitric acid																	
Mercurous sulphate paste																	

* Depending on the carbon.

The average temperature at the time these experiments were made was about 18° C. All the liquids and salts employed were chemically pure; the solids, however, were only commercially pure. The numbers in a vertical column below the name of a substance are the differences of potential, in volts, between that substance and the substance in the same horizontal row as the number, the two substances being in contact. Thus:—Lead is positive to distilled water, and the contact difference of potentials is 0·171 volts.

The authors point out that in all these experiments two air-contacts enter, and that up to the present time no direct experiment has enabled the difference of potential at *each* of these to be measured. They therefore show the importance of repeating their *quantitative* experiments on the electro-motive force of contact in other gases besides air, and especially in a very perfect vacuum; and they mention that, although they made the working drawings of the apparatus necessary for this extended investigation at the beginning of 1877, it was not until now that they have been enabled to commence it.

They further add:—" If the gas measurements such as we have indicated be extended to a good Crookes' vacuum, we may then possibly approximate to the real value of A B, the contact difference of potentials of A with B, the value in fact that we should obtain by a measurement of the Peltier effect.

RESULTS.

" The results which have been already obtained in this present investigation group themselves under three heads :—

" 1st. The contact difference of potentials of metals and liquids at the same temperature.

" 2nd. The contact difference of potentials of metals and liquids when one of the substances is at a different temperature from the other in contact with it; for example, mercury at 20° C. in contact with mercury at 40° C.

" 3rd. The contact difference of potentials of carbon and of platinum with water, and with weak and strong sulphuric acid.

" But those contained under head No. 1. are alone contained in the present paper." *

* The authors hope to have the honour of submitting the remainder of their completed experiments on a subsequent occasion to the Royal Society.

CHAPTER XX.

ACTIONS OF CURRENTS ON MAGNETS—COMMUTATORS—GALVANOMETERS.

ACTION ON MAGNETS.

LET a compass needle be suspended either on a point, or by a thread, and let a straight wire be brought near it, parallel to it, and above it (fig. 76). Let now a current be sent through

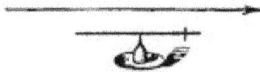
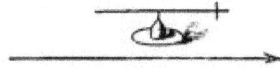

Fig. 76.   Fig. 77.

this wire. *The needle will be deflected, i. e. it will take up a position making a certain angle with the wire, which angle will increase as the strength of the current increases.* If the direction of the current be reversed—i. e. if the battery be turned round so as to bring the opposite poles to the ends of the wire—the direction of the deflection will be reversed. If the direction of the current be kept the same, and the wire placed below (fig. 77) instead of above the needle, the deflection will be reversed. If the direction of the current be reversed, *and* the wire placed below the needle, the deflection will be twice reversed, i. e. it will remain the same.

If now the wire be bent round the needle (fig. 78), the action of the bottom piece will cause the same deflection as that of the top one, for it is below the needle, and the current is in the

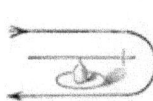

Fig. 78.   Fig. 79.

opposite direction, and so the action on the needle is greater than that of either portion singly. It may be bent again and again (fig. 79), and the effect is always to increase the deflection.

A current in a certain direction always deflects the marked end of the needle in the same direction. We will see which direction this is.

We suppose the current to flow from the positive to the negative pole of the battery.

Experiment will show us that if we imagine a little man swimming in the current with his face always turned to the needle, the north-pointing end of needle will always be deflected to his left hand.

An examination of the diagrams (Plate XXI.) will make this clearer than any explanation.

COMMUTATORS.

We have hitherto supposed that, when we change the direction of the current, we unscrew the wires and turn the battery round. In practice an instrument called a *commutator* is used to save doing this. It is made in many forms. We will only now describe some of the simplest, and allude to others when we come to experiments requiring them.

The simplest form (figs. 80, 81) consists of four thick quadrants of brass, screwed upon an ebonite base, so as just not to touch each other. Four holes are drilled at the division lines, each of which makes two nearly semicircular holes opposite each other. Into these, brass plugs can be put. When a plug is put in between any two quadrants, an electric current can pass from one to the other through the plug. Two plugs are used, and are both placed in one or the other diameter of the circle of quadrants. The two battery wires are attached to two diagonally opposite quadrants. The two ends of the wire through which we wish to send our current are attached to the two other quadrants.

Fig. 80. Elevation.

If now the plugs are placed as shown in fig. 81, the reader, if he will trace the course of the current, will see that it flows in the direction of the arrow, while, if the plugs are put in the other holes, it will flow in the opposite direction. If either plug is taken out, the current will cease altogether.

The next form (figs. 82, 83), though more complicated, is less

R

trouble to use, and, in using it, it is easier to remember in which direction the current is going.

On a mahogany base is an ivory roller supported on two brass

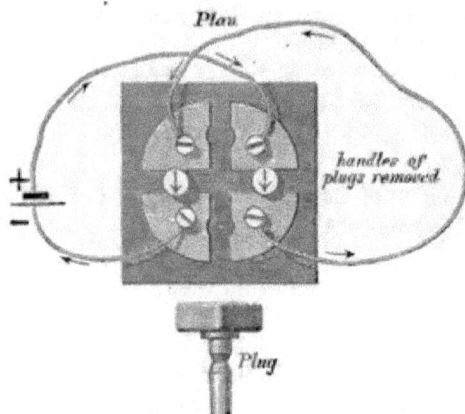

Fig. 81.

uprights. The axis of the roller is of brass, but is made in two pieces which do not meet in the middle. At opposite ends of a diameter of the roller, and on its circumference, are fixed pieces of

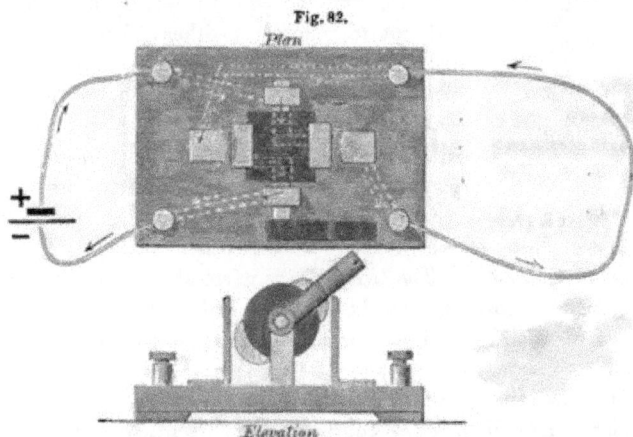

Fig. 82.

Fig. 83.

brass connected respectively to the two ends of the axis by brass pins inside the roller. Wires from the uprights pass under the base to two binding screws at one end of it. If now battery wires are fixed to these binding screws, then the brass plates fixed on the roller are in metallic communication with the poles of the battery respectively. An ivory handle, by which the roller can be turned half round, is attached to one end of the axis, parallel to the line joining the plates, so that when the handle is vertical, one plate is directly over the other. Fixed on the base, and coming up on each side of the roller, are two brass springs, which, by wires under the base, are attached to two other binding screws at the other end of it. To these latter are attached the ends of the wire through which the current is to be sent. We now see that when the handle is vertical, or nearly vertical, as in the elevation, no current can pass; but when it is horizontal, as in the plan, the metal plates on the roller press upon the springs, and a current passes in the direction of the arrow. If now the handle be turned so that it lies horizontally in the opposite direction, each plate will press on the spring opposite to that which it touched before, and the current will flow in the opposite direction. If the reader will trace the course of the current for himself on the plan, he will get a clearer notion of the working of the instrument than can be given by any explanation.

This commutator is a particularly convenient one for those branches of experimental work where alternate deflections have to be observed, for the battery wires can be so adjusted that the spot of light,* or the marked end of the needle, always travels in the direction in which the handle points. It is not, however, suitable for experiments where a sudden breaking of contact is required.

SPRING KEYS.—FIGS. 84, 85.

When a sudden make or break is required, "spring keys" are

Fig. 84.

Fig. 85.

used. They may be either simple contact keys, as fig. 84, or reversing keys with double springs, as fig. 85. In the latter, the

* See vol. i. page 253.

contacts are so arranged that, if one spring is pressed down, the current goes in one direction, while, if the other is pressed down, it goes in the other direction. Both keys are provided with a means of holding the springs in contact when required.

RAPID COMMUTATOR.—PLATE XXII.

In certain experiments it is required to reverse currents very rapidly. For this purpose the arrangement of Plate XXII. has been used. An ebonite frame carrying two wires of the shape oscillates above an ebonite plate, near the corners of which are four holes containing mercury. The battery wires are connected to the two oscillating wires respectively. The holes are joined diagonally, and each pair is connected to one end of the wire through which the current is to be sent. An examination of the plate will show that, when the ends b dip into the cups, the current will flow in one direction; and when the ends a dip, it will flow in the other direction. The oscillations are produced by a rod attached to a crank in the axis of a little electro-magnetic engine.* The speed is regulated by a friction brake, consisting of a loop of silk passed round a pulley on the axis, and attached to an india-rubber band, which can be tightened by turning a handle. This apparatus is useful for currents of high potential, and will easily give thirty reversals per second. It is the "secondary reversing engine" used by the present writer in his experiments in specific inductive capacity (vol. i. page 113). Numerous other forms of rapid commutators are used for different purposes.

GALVANOMETERS.

We have spoken of the different strengths of the currents produced by various batteries; we will now give an account of some of the methods used to measure them.

The instrument most commonly used to measure currents is called a "galvanometer."

Galvanometers may be roughly divided into two classes—

(1.) Those used to measure strong currents.

(2.) Those used to detect feeble currents.

The latter class are of great importance, for we shall see that most measurements of resistance are determined by a balancing of currents, such that, when the equilibrium is complete, no

* See Chapter XXVII.

PLATE XXII. — RAPID COMMUTATOR.

current shall pass; and, therefore, it is important to detect even the feeblest current.

The Tangent Galvanometer.

Among the first class the Tangent Galvanometer stands pre-eminent.

In its simplest form it consists of a large vertical ring of copper wire, in the centre of which is a small compass needle. When the instrument is used, it is turned so that the ring lies in the magnetic meridian, and therefore the needle lies in the plane of the ring. When an electric current is sent round the ring, it deflects the needle (page 240). Now, when the ring is large and the needle small, we can, in our calculations, neglect the difference of the distances of the various parts of it from the ring, i. e., we can consider the force acting on each part of it as equal to that acting at the centre of the ring.

The force f which a given current flowing in a circular arc exercises on a magnetic pole of strength m at its centre is equal to the strength of the pole multiplied by the strength of the current, multiplied by the length of the arc and divided by the square of the radius or distance of the wire from the magnet—

$$f = \frac{lm\,C}{a^2}$$

where l is the length of the arc and a the radius.

When the arc forms a complete ring the force is

$$\frac{2\pi a\,C^*}{a^2} = \frac{2\pi}{a}\,C$$

As soon as the current passes, a couple,[†] whose arm is the length of the compass needle, and whose forces are proportional to the strength of the current, begins to act on the needle.

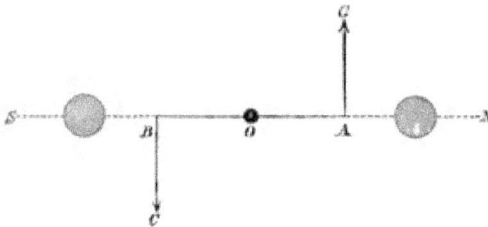

Fig. 86.

Let B O A (fig. 86) be the needle, and let its length be 2 l, and

* For the circumference of a circle is 2 π times the radius.
† See vol. i. p. 149.

let its pole be of strength m, so that the force exerted by any magnetic force on it is equal to m times that force; then, when first the current begins to act, we have a couple whose moment is $2\,lm\,\mathrm{C}\dfrac{2\,\pi}{a}$ *i. e.*, it is proportional to the length of the needle multiplied by the strength of the current. In these calculations it is, however, usual to consider only one half of the needle; then the moment of the couple acting on each half of the needle is $lm\,\mathrm{C}\dfrac{2\,\pi}{a}$ or half the whole couple.

Now the needle begins to move and is deflected through a certain angle, it may be 5°, 10°, or 50°—let us call it δ°, and then the same calculation will do for all experiments, and we can substitute the number of degrees observed in each experiment for δ in the final result.

Then the moment of the couple which the current exercises is less than before, for the direction of its force, which is perpendicular to the meridian, meets the meridian say at b (fig. 87).

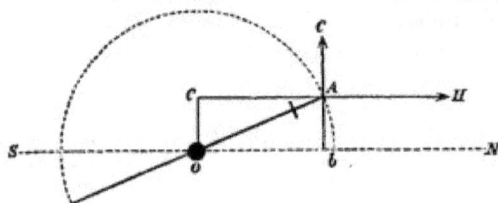

Fig. 87.

The arm of the couple is then no longer equal to O A, but only to Ob, and the moment of the couple is $\dfrac{2\,\pi}{a}\,m\,\mathrm{C}.\,\mathrm{O}b$, or is proportional to the strength of the current multiplied by the length Ob.

But, now that the needle is deflected, the earth's horizontal force, which we will call H, is acting in a direction parallel to the meridian, tending to pull it back into the meridian. This also forms a couple tending to turn the needle in the opposite way to the current. Its force is Hm, and its arm is evidently equal to the length Oc, that is to Ab; its moment is then Hm. Ab, or is proportional to the earth's horizontal force multiplied by the length Ab.

Now when the moments in opposite directions are equal, the

needle will come to rest, that is, when the needle is at rest th two couples are equal, i. e.

$$\frac{2\pi}{a} C \, m \, Ob = \text{H} \, m \, \Delta b,$$

which gives us—

$$\frac{2\pi}{a} C = \text{H} \, \frac{Ab}{Ob}$$

i. e. the strength of the current is proportional to the earth's horizontal force multiplied by the ratio of Ab to Ob.

We see that the length and strength of the needle makes no difference, for it cancels out. And it is known by trigonometry that the ratio of Ab to Ob depends only on the angle δ which the needle makes with the meridian, and not on the length of the needle or anything else.

This ratio is called the tangent of δ, and is always written tan δ.

We see that the ratio increases when δ increases. In books of mathematical tables the values of tan δ will be found, calculated for every value of δ, from $0°$ to $90°$.

We have now finally for the strength of the current—

$$C = \text{H} \tan \delta \, \frac{a}{2\pi}$$

—that is, the current equals the horizontal component of the earth's magnetism, multiplied by the tangent of the angle of deflection, multiplied by a constant which is determined by measuring the galvanometer ring.

Thus to find the strength of a current, we look in a magnetic chart for the strength of the earth's magnetism at the place. We then send the current through the galvanometer, and observe the angle of deflection, look for the number corresponding to that angle in a table of tangents, and, multiplying the two together ($\text{H} \tan \delta$), and by the constant $\frac{a}{2\pi}$, we have the strength of the current.*

---

* ACTION OF A CIRCULAR CURRENT ON A COMPASS NEEDLE AT ITS CENTRE.—FIG. 88.

General Case:—

Let the compass needle be free to move only in a horizontal plane.

Let the plane of the ring make an angle θ with the vertical, and let a be the horizontal angle through which the needle would have to be turned to bring it from the magnetic meridian into the plane of the ring.

Let C be the current, H the horizontal component of the earth's force, a the radius of the ring, δ the angle which the needle makes with the meridian when at rest under the joint actions of the earth and current.

The single ring is the simplest form of the galvanometer, but it is not suited to accurate work, owing partly to the want of security as to the ring remaining perfectly plane and rigid, and partly to the comparatively high ratio which the irregular action of the connecting wires bears to the regular action of the ring.

Then the couple tending to bring the needle to zero is
$$H \sin \delta \, lm.$$

The couple tending to deflect it is
$$\frac{2\pi}{a} C \cos \theta \cos (\text{angle between ring and needle}) \, lm$$

$$= \frac{2\pi}{a} C \cos \theta \cos (\delta \pm a) \, lm$$

(the sign of a depending on the direction of the current)

$$= \frac{2\pi}{a} C \cos \theta \{\cos \delta \cos a \mp \sin \delta \sin a\} \, lm$$

$$\therefore C = H \frac{\sin \delta}{\cos \theta \{\cos \delta \cos a + \sin \delta \sin a} \cdot \frac{a}{2\pi}$$

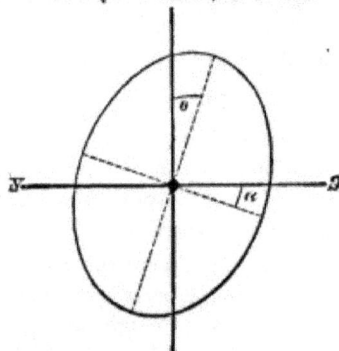

Fig. 88.

Particular Cases :—

(1) Ring horizontal—
$$\theta = \frac{\pi}{2} \qquad \cos \theta = 0$$
makes $\qquad C = \infty$ except when $\delta = 0$
or, it would require an infinitely strong current to produce any deflection.

(2) Ring vertical—
$$\theta = 0 \qquad \cos \theta = 1$$
makes $\qquad C = H \dfrac{\sin \delta}{\cos \delta \cos a + \sin \delta \sin a} \cdot \dfrac{a}{2\pi}$

(3) Ring vertical, and in the magnetic meridian (tangent galvanometer)—
$$\theta = 0 \qquad a = 0$$
$$\cos \theta = 1 \qquad \cos a = 1 \qquad \sin a = 0$$
makes $\qquad C = H \dfrac{\sin \delta}{\cos \delta} \dfrac{a}{2\pi} = H \tan \delta \dfrac{a}{2\pi}.$

The following is a better form of the instrument (fig. 89):—

Two rings are used, one on each side of the needle, so placed that the centre of the needle is at the centre of their common

Fig. 89.

axis (or line joining their centres), and that this line, when the instrument is adjusted, is at right angles to the magnetic meridian, i. e., lies magnetic E. and W. The rings are made of wood, and the wires are wound on them. The section of a ring is as in fig. 90. In a groove at the outside is a massive copper ring, which is used only for rough experiments with very powerful currents. The rings on each side are connected so

---

(4) Ring in the magnetic meridian but not vertical—

$$a = 0 \qquad \cos a = 1 \qquad \sin a = 0$$
$$C = H \frac{\sin \delta}{\cos \theta \cos \delta} \cdot \frac{a}{2\pi} = H \frac{\tan \delta}{\cos \theta} \frac{a}{2\pi}$$

A galvanometer, of which the ring can be turned round a horizontal axis so as to make a measured angle θ with the vertical, has been constructed for the measurement of very powerful currents, and is called the cosine galvanometer. The above formula is used with it. With a given current the deflection can be made as small as we please by increasing θ.

(5) Ring vertical, and east and west—

$$a = \frac{\pi}{2} \qquad\qquad \theta = 0$$
$$\cos a = 0 \qquad \sin a = 1 \qquad \cos \theta = 1$$

that the current goes in the same direction through each. The inside of each ring is turned so as to make part of a cone, such that, if it were continued, its apex would be at the centre of the needle. On this three coils of wire, of respectively 3, 9, and 27 windings, are laid; they are, of course, covered with silk or cotton to make the current go round and round, instead of merely across from one wire to another. Each winding produces its own effect on the magnet, and thus with a current, such that its effect when

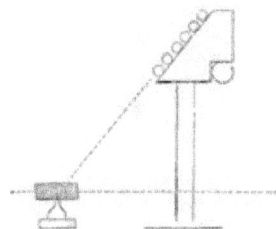

Fig. 90.   Fig. 91.

in the single ring is unity, we can produce an effect on the needle equal to 3, 9, or 27, or any combination of the sums or differences of those numbers.

When there are n windings, the equation for the tangent galvanometer becomes

$$C = H \tan \frac{a}{2 n \pi}.$$

The reason for winding on the cone is that the solid angle subtended at the needle by each winding is the same, which eliminates the error caused by the needle not being indefinitely short.

The needle is a short pointed one with a piece of agate let in to the top of the cap where it rests on the pin, while to allow a larger divided circle to be used, a light aluminium needle is attached to it (fig. 91).

The needle is arranged so that the points of the aluminium needle are as nearly as possible in the line of magnetization.

---

makes $$C = H \frac{\sin \delta}{\sin \delta} \frac{a}{2 \pi}$$

But in this case there can be no deflection, as C acts in the same direction as H; and, therefore,

$$\delta = 0 \text{ and } \sin \delta = 0;$$

and $$C = H \frac{0}{0}$$

which is indeterminate; or, with the ring in this position, the needle gives no information about the strength of the current.

Any error in this adjustment is, however, corrected by taking alternate readings with the current in opposite directions; the one reading will then be as much too great as the other is too small, and the mean will be the true reading.

Let us take an extreme case as an illustration, and suppose that the angle between the magnetic axis of the needle and the line joining the points is 1°, and let us suppose that the current causes a deflection of 30° of the magnetic axis (fig. 92). Then the reading in one direction will be 29°, and in the other 31°. The mean is 30°, the true deflection; and we see that to determine this does not require a knowledge of the angle between the direction of the pointer and the magnetic axis of the needle. This can, however, be easily obtained if wanted, for it is evidently half the difference between the two readings. (Compare vol. i. page 179.)

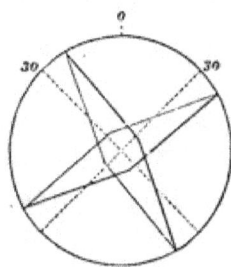

Fig. 92.

The instrument is supported on three levelling screws, and the base and supporting pillar are connected by a pivot, which enables the latter to be turned round so as to adjust the circles in the meridian.

SENSITIVE GALVANOMETERS.—ASTATIC NEEDLES.

Under the head of *sensitive galvanometers* come all those used for detecting or measuring feeble currents. The *astatic needle* is an arrangement used in most galvanometers of this sort to diminish the earth's couple, while at the same time it slightly increases the couple of the current. The needle consists of two magnets, almost, but not exactly, of the same strength, connected together by a rigid bar, with their similar poles in opposite directions. They are not pivoted, but are hung by a silk thread. The marked end of the stronger magnet will point to the north, but if the combination be deflected by any means, the couple tending to bring it back to the meridian will only be the difference of the couples exerted by the earth on the two magnets respectively. The coil of wire through which the current passes has an opening left near the centre of the top side, and the connecting bar of the magnets passes through it. One magnet thus hangs inside the coil, and the other just above it (fig. 93). A reference to Plate XXI. will show that the actions of the top and bottom of the coil on the lower needle are in the same direction, while, though the actions on the upper needle are in opposite directions, that of

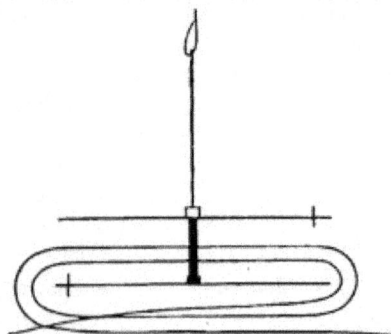

Fig. 93.

the top of the coil which is nearest, and therefore most powerful, is in the same direction as those on the lower one.

In sensitive galvanometers the current goes many, often several thousand, times round the needle. The wire is wound on a flat reel of the form shown in fig. 94. A little tube a is fixed to it, so as to leave an opening for the axis of the needles.

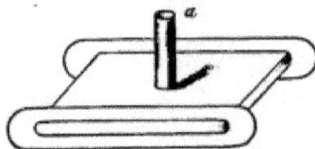

Fig. 94.

Fig. 95.

The indicator, which moves over a divided circle, is of course attached to the top needle, which can be seen. The ordinary form of the astatic galvanometer is shown in fig. 95.

We see it consists of an astatic needle suspended in a coil of wire, which, according to the purpose for which the galvanometer

PLATE XXIII.—THOMSON'S REFLECTING GALVANOMETER.

is required, consists either of a few turns of thick, or of many turns of thin, wire. The needle is suspended by a fibre of unspun silk attached to a brass support. A glass shade protects it from currents of air. Through a hole in the top of the glass shade projects the top of a sliding piece, to which, and not directly to the fixed support, the upper end of the fibre is attached. This enables the needles to be lowered upon the coils, so as to take the strain off the silk fibre when moving the instrument. A divided circle on a card is fixed to the coil, and circle and coil can be turned through 30° or 40° on a pivot, so as to adjust them to the meridian of the needle. Sometimes a mirror like that described for the reflecting electrometer is fixed to the needle, and allows the galvanometer to be used with a lamp and scale. This is particularly useful for lecture purposes.

With the aid of a lime light, the author has made the movements of a very small needle visible to 500 people at once. A disc of light six inches in diameter was thrown on a screen, and a deflection of 1° or 2° moved it several feet. The needle was about two inches long, and the mirror ¼ in. in diameter. But when a mirror galvanometer is required for accurate work, one of Sir Wm. Thomson's forms is always used.

Sir Wm. Thomson's Reflecting Galvanometers.

Plate XXIII.

The optical principle of these galvanometers is precisely the same as that of the reflecting electrometer (vol. i. page 38). These

galvanometers are sometimes made astatic, sometimes not. The mirror is usually less than ¼ in. in diameter, and is very thin. In the nonastatic form (fig. 96) the magnet, or rather magnets, for several are generally used, are cemented to the back of the mirror, and are usually about ⅛ in. in length. The whole system of magnets and mirror weighs less than a grain. The object of having several magnets is to get the maximum of magnetization* with the minimum of weight. The mirror and magnets are hung by a single fibre of unspun

Fig. 96.

* This is required because the more highly the needle is magnetized the more rapidly will it come to rest after being set swinging.

silk in the centre of a circular coil of wire, which is enclosed in a brass cylinder. The front end of the cylinder is closed by a glass plate, the back by a brass one, in the centre of which is a small disc of plate glass, through which the mirror can be seen. The cylinder is supported horizontally on a tripod stand, each leg of which is furnished with a levelling screw, by which the apparatus is adjusted until the mirror is seen to swing clear.

To avoid the inconvenience of having to place the apparatus always in the magnetic meridian, a large curved magnet feebly magnetized is supported horizontally on a vertical stem, fixed to the top of the case.

The magnet can be turned by hand on the bar on which it slides somewhat stiffly; and by its directive force makes an artificial magnetic meridian in any desired direction. A fine adjustment is obtained by moving the stem itself by means of a tangent screw. The magnet can also be slid up or down the stem so as to act more or less powerfully on the suspended magnet.

The scale and lamp arrangement (fig. 13, vol. i. page 38) is precisely the same as that used with the electrometer already described, except that instead of a wide opening with a line across it, a narrow slit making a bright vertical line of light is generally used. This galvanometer is made sometimes with a short thick wire, sometimes with a long thin one.

In the astatic form (fig. 97 and Plate XXIII.), which is used only with long wire galvanometers, each needle is surrounded by its own coil of wire. The current of course goes opposite ways in these two coils. The coils are sometimes enclosed in a vertical cylinder of glass, as in fig. 97, and sometimes in a square

Fig. 97.

glass case, as in Plate XXIII. As the magnet and mirror system is necessarily somewhat heavier in this construction, an aluminium fan is sometimes attached to it to check its vibrations. The other details are similar to those of the tripod form.

The Marine Galvanometer.—Fig. 98.

This instrument is used on board ship in laying telegraph cables. Its needle, instead of being suspended, is strung on a fibre secured at top and bottom. A large permanent magnet is fixed near it in the case, so that the meridian has a sensibly constant direction with regard to the case. For a fine adjustment two sliding magnets with their opposite poles turned towards

Fig. 98.

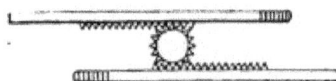

Fig. 99.

the needle, are connected by a cog-wheel (fig. 99). When the poles are equidistant from the needle, they produce no effect, but, by turning the wheel to the right or left, one or the other can be made to preponderate. They are contained in the horizontal tube seen on the left of fig. 98. Except the small window through which the mirror is observed, the whole of the case of the instrument is of iron, which almost entirely destroys the effects of external magnetic forces.

Balistic Galvanometer.

When a quantity of electricity is instantaneously discharged through a galvanometer, the quantity can be calculated from the limits of the first swing of the needle.

It can be shown mathematically that the quantity of electricity which passes is, if the air offers no resistance to the motion of the needle, proportional to the sine of half the angle of swing.*

In order to diminish the resistance of the air as much as possible, a "*balistic galvanometer*" has been used.

* Maxwell, "Electricity," 748, vol. ii. p. 346.

The difference between a balistic and an ordinary galvanometer is this:—In the former, it is desired to bring the needle to rest as soon as possible; in the latter, where the limit of swing is the quantity to be observed, it is desired to check or "damp" the swing as little as possible.

The following form of the instrument has been contrived by Professors Ayrton and Perry.[*]

One of Elliott's high-resistance reflecting galvanometers (Plate XXIII.) was used, but the needles were removed and replaced by the following arrangement :—

Forty small magnets of varying lengths were prepared, and, having been magnetized to saturation, were built up into two little spheres, in each of which all the magnets pointed in the same direction. The spheres were completed by segments cut from a small hollow leaden ball. The two spheres were rigidly connected so as to form an astatic combination, which was suspended in the ordinary manner.

With this arrangement great sensibility was obtained, and the air offered very little resistance to the motion of the needles. It was found that the ratio of the first swing to the second was only 1·1695, a number which is sufficiently near to unity to allow a simple correction to be applied for the damping effect of the air.

DIFFERENTIAL GALVANOMETER.

For comparing the strength of two currents galvanometers are sometimes made with two exactly equal coils of wire, round which the currents can be sent in opposite directions. When the currents are equal there is no deflection of the needle.

Such an instrument is called a " Differential Galvanometer."

[*] Report of the British Association, Dublin, 1878, page 487, and Phil. Mag., 1879, i. p. 277.

CHAPTER XXI.

ELECTRIC RESISTANCE AND ELECTRO-MAGNETIC UNITS.

No bodies are perfect conductors of electricity; all offer resistance to its passage.

OHM'S LAW.

Ohm's Law, which is the foundation of modern electrical measurement, is this—

The strength of the current in a wire or other conductor is directly proportional to the difference of potential between its ends, and inversely proportional to its resistance.

The difference of potential at the ends is usually called " the electro-motive force." The identity of the two expressions is obvious from our definitions on page 26 of this volume. We may thus say that the strength of the current equals the electro-motive force divided by the resistance.

Briefly, if C be the strength of a current, E the electro-motive force, and r the resistance of the circuit, we have

$$C = \frac{E}{r} \quad . \quad . \quad . \quad . \quad (1)$$

or

$$E = Cr \quad . \quad . \quad . \quad . \quad (2)$$

i. e., electro-motive force equals current multiplied by resistance—
or

$$r = \frac{E}{C} \quad . \quad . \quad . \quad . \quad (3)$$

i. e., resistance equals electro-motive force divided by current.

ELECTRO-STATIC MEASURE OF CURRENT AND RESISTANCE.

If we have two conductors at different potentials, and connect them by a wire, there will be a flow of electricity or a current

s

along the wire; and if the electro-static unit of quantity, as defined on page 53 of this volume, passes in each second, the strength of the current (in electro-static measure) is said to be equal to unity.

Now let us call Q the quantity of electricity, as defined on page 53, conveyed by a current C in a time t. Then Q equals the strength of the current, multiplied by the time for which it lasts, i.e.,

$$Q = Ct \quad . \quad . \quad . \quad . \quad (4).$$

In the same way the quantity of water which flows through a tap equals the strength of the stream multiplied by the time for which the tap is left open.

By Ohm's law the unit of electric resistance is in electro-static measure such that, with any number of electro-static units of difference of potential, the same number of electro-static units of electricity will pass through the resistance in one second; for we see by equation (3) that r is unity when $E = C$.

Also from (1) and (4) we have

$$Q = \frac{E\,t}{r} \quad . \quad . \quad . \quad . \quad (5)$$

or

$$r = \frac{E\,t}{Q} \quad . \quad . \quad . \quad . \quad (6)$$

Hence, the resistance of a conductor,[*] measured in electro-static measure, is equal to the time required for the passage of a unit of electricity through it when unit difference of potential is maintained between its ends: and—

$$\text{Resistance} = \frac{\text{Time} \times \text{Electro-motive force}}{\text{Quantity}}.[†]$$

ELECTRO-MAGNETIC MEASURES.

Before studying the electro-magnetic measure of resistance it is necessary to consider the electro-magnetic measures of quantity and potential.

ELECTRO-MAGNETIC MEASURE OF CURRENT.

In the description of the tangent galvanometer, we found that the strength of the current, as measured by the tangent galvanometer, is given by the following equation :—

$$C = H \tan \delta \frac{a}{2\,n\,\pi}.$$

[*] For B.A. unit of resistance, see Chap. XXVII.
[†] Everett, " Units and Physical Constants," p. 130.

The unit of current strength is got by putting $C = 1$ in the equation,

$$C = H \tan \delta \frac{a}{2 n \pi}$$

—that is, it is the current which will give a deflection δ when

$$\tan \delta = \frac{2 n \pi}{a H}.$$

Problem.—Assuming the earth's horizontal force to have its present average value at Greenwich of ·1794, what deflection would a unit current*produce in a single ring tangent galvanometer with a ring $\frac{1}{2}$ a metre diameter?

We have $n = 1$, $\pi = 3·1416$, $H = ·1794$, $a = 25$ centims.

Hence,

$$\tan \delta = \frac{2 \times 3·1416}{25 \times ·1794}$$

which corresponds to an angle of deflection of

$$\delta = 54° \; 6'.$$

ELECTRO-MAGNETIC MEASURE OF QUANTITY.

The quantity Q conveyed by a current in a given time is the product of the current by the time which it lasts.

The electro-magnetic unit of quantity is the quantity of electricity that a unit of current conveys in a unit of time—that is, in a second.

ELECTRO-MAGNETIC MEASURE OF POTENTIAL, OR ELECTRO-MOTIVE FORCE.

The work done by an electro-motive force E, in urging a quantity of electricity Q through a conductor, is E Q—that is, the work equals the quantity of electricity multiplied by the electro-motive force.

The unit electro-motive force is that which must be maintained at the ends of a circuit, that unit quantity of electricity may de a unit of work in flowing through it.

ELECTRO-MAGNETIC MEASURE OF RESISTANCE.

With the help of Ohm's law we can now define the electro-magnetic unit of resistance.

The unit of resistance is an electro-motive force divided by a current numerically equal to it; for we have $r = \dfrac{E}{C}$, and $r = 1$

* A "unit current" equals 10 ampères.

s 2

when $E = C$; so that a wire is said to have unit resistance when any electro-motive force acting along it produces a current numerically equal to itself.

When we come to the theory of dimensions we shall be able to show that this resistance is a velocity—that is, that the ratio of electro-motive force to current is the ratio of a length to a time.

ELECTRO-MAGNETIC MEASURE OF CAPACITY.

The condenser whose capacity is unity is that which will be at unit potential when charged with unit quantity, or, in other words, it is the condenser whose potential will be always equal to its charge, both being expressed in electro-magnetic measure.

PRACTICAL UNITS—VOLT, OHM, AMPÈRE, COULOMB, FARAD.

As some of these units are of inconvenient magnitudes, certain decimal multiples and sub-multiples of them are used in practical work.

ELECTRO-MOTIVE FORCE—VOLT.

The practical unit of electro-motive force is called the *Volt*, and is equal to one hundred million absolute units of potential.

$$1 \text{ Volt} = 10^8 \text{ C.G.S. units.}$$

A Latimer Clark's * standard cell has an electro-motive force at $15°\cdot5$ C. of $1\cdot457$ Volt.

RESISTANCE—OHM.

The practical unit of Resistance is called the *Ohm*, and is equal to one thousand million absolute electro-magnetic units.

$$1 \text{ Ohm} = 10^9 \text{ C.G.S. units.}$$

The standard Ohm is the resistance of a column of mercury at a temperature $0°$ C. of one square millimetre section, and whose length is immediately to be accurately determined by an international commission appointed at the Congress of Electricians which met in Paris on Sept. 15, 1881. The length is known to lie between $104\frac{1}{2}$ and $105\frac{1}{2}$ centims. See page 323.

CURRENT—AMPÈRE.

The practical unit of current is called the *Ampère*, and equals $\frac{1}{10}$ the C.G.S. unit.

* Vol. i. page 220.

QUANTITY—COULOMB.

The practical unit of quantity is called the *Coulomb*. It is equal to $\frac{1}{10}$ the absolute electro-magnetic unit of quantity.

1 Coulomb = $\frac{1}{10}$ C.G.S. unit of quantity.

CAPACITY—FARAD.

The practical unit of capacity is called the *Farad*. It is equal to the one-thousand-millionth part of the absolute electro-magnetic unit.

1 Farad = 10^{-9} C.G.S. units.

As the Farad is too large for practical use, condensers are constructed, each having a capacity of one-millionth of a Farad. This capacity is called a *Microfarad*.

Fig. 100.

A condenser of one Microfarad capacity would contain *about* 300 circular sheets of tinfoil separated by mica plates, and would be contained in a box (fig. 100) 3½ inches deep, and 6½ inches diameter.*

RELATIONS BETWEEN THE PRACTICAL UNITS.

The Ampère, or practical unit of current, is the current produced by an electro-motive force of one Volt working through a resistance of one Ohm.

The Coulomb, or practical unit of quantity, is the quantity of electricity conveyed in one second by a current of one Ampère.

The Farad is the capacity of a condenser which holds one Coulomb at a potential of one Volt.

The Microfarad is the millionth part of a Farad.

These practical units, the relations between them, and the names to be given to them, were defined and settled at the International Congress of Electricians mentioned above.

* The insertion of the plug connects the two coatings and discharges the condenser.

CHAPTER XXII.

It is found by experiment that, with any given material in a homogeneous state, the resistance of a wire of uniform section varies directly as its length.

The resistance of a wire of given length varies inversely as its cross section—that is, inversely as its weight.

Therefore, the resistance of a uniform wire of given material is proportional to its length, divided by its weight.

The resistance of different materials varies. For instance, if that of copper be represented by 1615, that of iron would be 9827.

It is obvious that if any arbitrary wire be taken as a standard, and the resistances of other wires compared with it, the numbers thus obtained will be proportional, but not in general equal, to their resistances expressed in absolute measure.

Certain wires, called resistance coils, are prepared by a method which will be described later, so that their resistances may be known in absolute measure, and then all other wires are compared with them.

WHEATSTONE'S BRIDGE.

We will suppose that we are furnished with a set of resistance coils, and that a wire is given to us of which we are to determine the resistance. We use an arrangement invented by Mr. Christie, and called "Wheatstone's Bridge."

We will first describe what may be called the "lecture model" of the apparatus, as it is a machine easy to understand, but inconvenient to work with, and then go on to describe the forms in practical use.

The lecture model (fig. 101) consists of a board on which is fixed a "diamond" of metal strips. At the four corners, A B C D, are binding screws, while in each side is a break with binding screws at each end of it. To two corners opposite to each other are connected the battery wires; to the two other corners, those of the galvanometer. In the four breaks are put three known resistances, which we call S, *s*, and R, and the unknown one which we call *x*. We then vary the resistance R, and we shall find that at .one particular value, the current of the battery produces no deflection of the galvanometer. When this is the case, we have, as we will prove immediately,

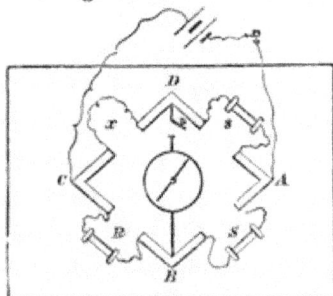

Fig. 101.

Ratio of *x* to R equals ratio of *s* to S;

from which *x* can be found by simple proportion.

THEORY OF WHEATSTONE'S BRIDGE.

To understand this, we must note the following direct deduction from Ohm's law :—

If a wire of uniform resistance be connected to a battery, the potential varies regularly from one end to the other of the length— that is, at the middle point the difference of the potential from that at either end is half that of the ends. At $\frac{1}{3}$ from one end the potential differs from the potential at that end by $\frac{1}{3}$ of the whole difference ; and so on.

More generally in any wire, the potential varies regularly along the resistance—that is, if there be a wire of 10 units resistance, and the potential at one end is zero, and at the other is 100, the potential at one unit from the first end will be 10, at two will be 20, at three 30; and so on.

When the battery is in action, the current, on arriving at A (fig. 99), divides, as a stream might divide into two channels round an island, and part goes by the road ADC; part by ABC.

Let us now draw straight lines, ADC, ABC (figs. 102, 103), representing the resistances in the two courses, and let us draw vertical lines AL at the ends A, representing the difference of potential between A and C.

Let AD represent the resistance *s*, DC the resistance *x*, then the total resistance of the branch of the circuit, containing *s* and *x*, is represented by the line ADC. Similarly the resistance of the branch containing S and R is represented by ABC.

The length of the line AL, which represents the excess of

the potential at A over that at C, is of course the same in both diagrams. Draw lines LC in each. Now, by what we have just stated—viz., that the potential diminishes regularly—the potential* at any other point in the circuit can be represented by the length of a vertical line drawn from the horizontal line at that point to the sloping line. The potential at D, where one of the galvanometer wires is attached, is represented by the length DM (fig. 102).† In a similar way that at B, where the other wire is attached, is represented by BN (fig. 103).

Now, the effect of altering the resistance R will be to alter

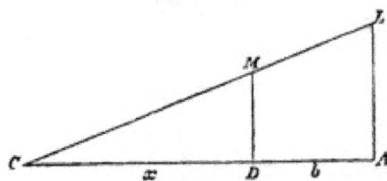

Fig. 102.

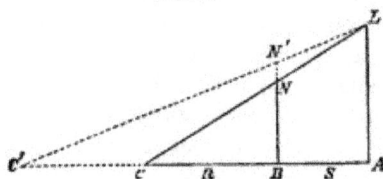

Fig. 103.

the potential at B; for, suppose R increased so as to bring C to the position C', then the fall of the potential would be represented by the dotted line LC', and the potential at B would be represented by the length BN'. Let us then vary R until BN equals DM, *then, the ends of the galvanometer wire are at the same potential, and there is no current through it.*

But as the height AL is the same in both triangles, the heights at any other points D and B in the bases respectively can only be equal when the ratio of **AD to DC**—that is, of *s* to *x*, is equal to the ratio of **AB to BC**—that is, of S to R‡—or, the galvanometer is at zero when

Ratio of *s* to *x* equals ratio of S to R.

But when two ratios are equal, the ratio of the first term of

---

* By " potential " we mean " excess of potential over that at C."
† The galvanometer circuit being broken.
‡ The student is advised to test this with a scale drawing.

one to the first term of the other is equal to the ratio of the second term of the one to the second term of the other; therefore, when the galvanometer is at zero, we have

Ratio of x to R equals ratio of s to S;

or, x equals R, multiplied by the

Ratio of s to S,

which is written—

$$x = R\frac{s}{S};$$

Thus, when the galvanometer is at zero, x is known if the other three quantities are known.

The student will find it a useful exercise to prove for himself that it is unimportant at which pair of corners the battery wires are attached—that is, that interchanging the battery and galvanometer makes no difference.

In practice the galvanometer and battery should be so arranged

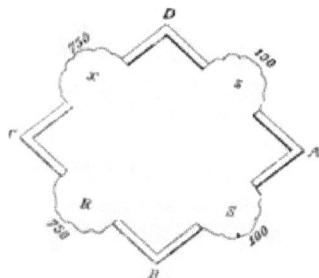
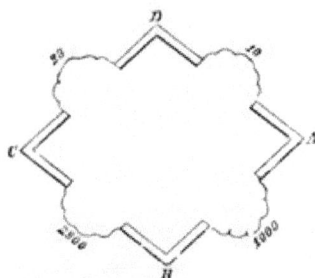

Fig. 104.   Fig. 105.

that the two branches of the battery current encounter as nearly as possible the same resistance.

For instance, let us suppose S and s each equal 100, and R equal 750. Then $\frac{s}{S} = 1$; and, therefore, $x = R$, and our four branches will be as in fig. 104. The battery wires should now be attached to AC, and the resistance in each branch will equal 850. If they are attached to BD, the resistance in the one branch is 1500, and that in the other only 200. Now, however, suppose we have

S = 1000, s = 10, and R = 2800.

We have, to find x,

$$x = R\frac{s}{S} = 2800\,\frac{10}{1000} = 28.$$

We must now attach the battery wires to BD, as in this case the resistances of the two branches will be, as in fig. 105, 1010 and 2828; whereas, if we attach them to AC, the circuits will be 38 and 3800.

The objection to having a large difference in the branches is that nearly the whole current then passes through the circuit of least resistance, and heats it, thereby increasing the resistance on that side (for the resistance of a wire increases when it is heated), while, when the current is about equally divided, the increase of resistance due to the heating is about the same on both sides.

For convenience of calculation it is usual to make S a decimal multiple, or submultiple of s; then, when the balance is established by varying R, it is only necessary to multiply R by some power of 10, to find x.

PRACTICAL FORMS OF WHEATSTONE'S BRIDGE.

The form of bridge above described is never used except for lecture purposes. Two principal forms are used in practice. The one which is most used is called a "resistance box," and the other is called the "sliding bridge."

SLIDING BRIDGE.

This latter (figs. 106, 107) consists of a horizontal board, with a

Fig. 106.

straight wire of high resistance stretched along one side, and a copper strip with gaps in it along the other. The diagram

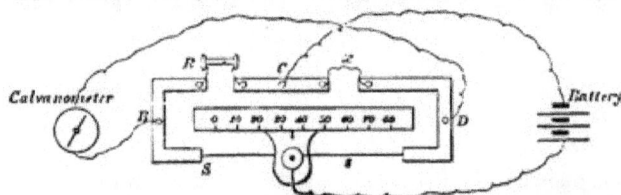

Fig. 107.

(fig. 107) should be compared with that of the lecture model (fig. 101)—the same letters indicate the same points in both.

The connections are made as shown ; R is not variable, but is

Standard Unit of Electrical Resistance.

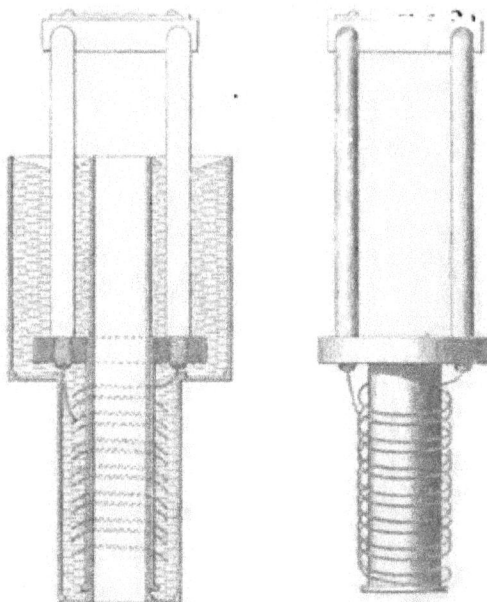

SECTION.

ELEVATION.

$\frac{1}{2}$ full size.

Metal.

Ebonite.

Parafine.

Solder.

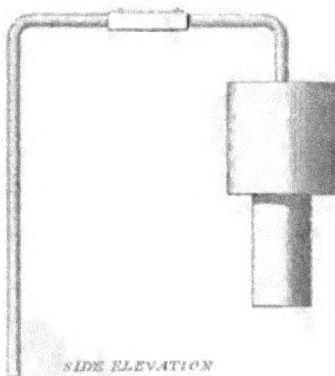

SIDE ELEVATION

PLAN.

$\frac{1}{4}$ full size.

Plate XXIV.

chosen arbitrarily. A, one of the battery wires, is attached to a slider which slides along the resistance wire. Now the wire at one side of A is S, that at the other *s*. It is obvious that, *if the resistance of the wire is uniform*, the ratio $\frac{s}{S}$ will be equal to the ratio of the lengths of the wires on each side of the slider. A scale is fixed on the base, and if we suppose the equilibrium to be established with the slider at, say, 35, as in fig. (107), and suppose that R = 100, we shall have

$$x = \text{R}\frac{s}{S} = 100\,\frac{35}{45} = \frac{700}{9} = 77.\ \&c.$$

We see in this apparatus that it is not necessary to know the values of S and *s*, but only their ratio.

The apparatus is useful for some purposes, but it is not susceptible of any great accuracy, as its working depends on the assumption that a wire exposed to the air has a uniform resistance. As every particle of rust or scale, formed or rubbed off, and every scratch made by the slider affects the resistance, we see that the assumption cannot be held to be strictly true.

The slider usually carries a spring and vertical sliding rod, to which latter the battery wire is attached, so that contact is only made when wanted, by pressing down the spring.

RESISTANCE COILS.—PLATE XXIV.

A coil of wire of a known resistance is called a *Resistance Coil.* Resistance coils are usually made either of German silver wire, or of an alloy of silver with 33·4 per cent of platinum, as the resistance of those materials varies very slightly with changes of temperature.

The two ends being fixed to massive copper rods, the wire, previously carefully insulated with two or more layers of silk, is wound double upon a reel, as shown in Plate XXIV. It is usual to enclose the reel in a thin brass case, and imbed the wire in paraffin. By immersing the case in water, the wire can be brought to any desired temperature. The effect of the double winding is that there are always two equal currents in opposite directions close together. This entirely prevents any inductive effect on neighbouring magnets or wires.

BRIDGE RESISTANCE BOX.

A *resistance box* consists of a number of coils of different resistances arranged in the following manner. The coils are all

fixed to the under side of a slab of ebonite, which forms the lid
of a mahogany box (fig. 108). On the top of the lid are a number

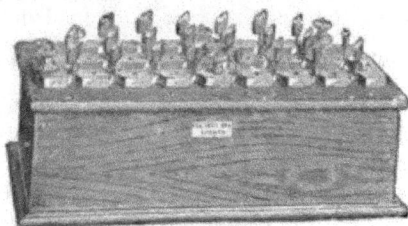

Fig. 108.

of brass blocks, to each of which one end of each of two coils is
connected, as shown in section in fig. 109. Thus, if a current is

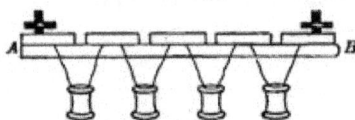

Fig. 109.

sent from A to B, it has to pass through all the coils. The ends
of the brass blocks are, however, shaped as shown in the plan
(fig. 108), and brass plugs * fit in between them. When a plug is
put in at any opening (*a*, fig. 110), the current only passes through

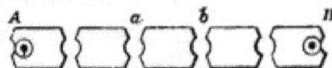

Fig. 110.

the plug, and not by the coil; so that when a plug is put in, say at
a, the total resistance is less by the resistance of the coil at *a*,
or, generally, the resistance from A to B is equal to the resistance
unplugged. The resistance of each coil is engraved on the lid
near the plug-hole. Here, then, we have a means of varying the
resistances without interfering with the connections.

The coils in the bridge resistance box are arranged as in fig.
111, the numbers representing the number of units of resistance
in each coil.

The reader is again requested to compare this figure with
the picture of the lecture model, fig. 101, p. 263. The same
letters are used for the same points.

We see that the coils are arranged in a continuous line, and
binding screws inserted at certain intervals.

* Similar to the plug shown in fig. 81, vol. i. p. 242.

The first line contains two sets each of 10, 100, and 1000.

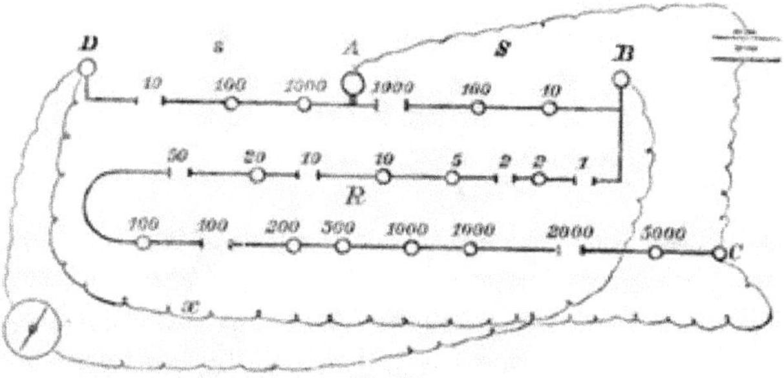

Fig. 111.

These form the branches S and *s*. We see that they give the following values of $\frac{s}{S}$:—

$$\frac{10}{1000} = \frac{1}{100} \qquad \text{also}$$

$$\left.\begin{array}{c}\frac{10}{100} \\[4pt] \frac{100}{1000}\end{array}\right\} = \frac{1}{10} \qquad \frac{1010}{100}$$

$$\left.\begin{array}{c}\frac{10}{10} \\[4pt] \frac{100}{100} \\[4pt] \frac{1000}{1000}\end{array}\right\} = 1 \qquad \frac{1100}{10}$$

$$\left.\begin{array}{c}\frac{100}{10} \\[4pt] \frac{1000}{100}\end{array}\right\} = 10 \qquad \begin{array}{c}\frac{10}{1100} \\[4pt] \frac{100}{1010}\end{array}$$

$$\frac{1000}{10} = 100$$

Different forms of the same fraction are used for different resistances.

Thus, if we desired to have $\frac{s}{S} = 1$ we should make it $\frac{1000}{1000}$ if *x* were large, $\frac{10}{10}$ if it were small.

The arrangement of the bridge has been drawn in the way most convenient for calculation; but if the resistance were such that there was equilibrium with the present resistance unplugged, it would be better to interchange battery and galvanometer, as at present the two currents are inversely, as

$$10 + x \text{ and } 1000 + 2163.$$

Now

$$x \text{ is } R\frac{s}{S} = 2163 \ \frac{10}{1000} = 21\cdot63,$$

and therefore we have

$$31\cdot63 \text{ and } 3163;$$

whereas, after interchanging, we have

$$1100 \text{ and } 2184\cdot63.$$

The longer part of the line of coils represents R on the bridge, and the arrangement of it is worth notice. There are only sixteen coils, and yet, by combining them, any number from 1 to 10,000 can be obtained. This is best seen by taking any number at random and trying to make it up. Generally speaking, we must use the large numbers first—that is, we must first unplug the largest number below the number we want.

x, the wire whose resistance is to be found, is attached to the binding screws C and D. If it should not be long enough to reach from one to the other, the length must be made up either by a wire of known resistance, or better by a copper rod so thick that its resistance may be neglected.

ATTACHMENT OF THIN WIRE.

Where x is a thin wire, it should not be attached directly to the binding screws, as there would be an uncertainty as to the exact point of it in contact, and therefore as to the length between the screws. It should instead be soldered to a piece of thick copper wire beaten into the shape shown in fig. 112.

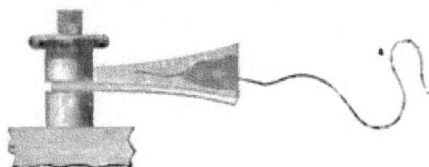

Fig. 112.

The binding screws of C and D are each made with double nuts, so that different wires can be tested without disturbing the galvanometer connections.

CONTACT KEYS.

Contact keys are placed in both the galvanometer and the battery circuits; that in the battery circuit may be of any form so long as it enables us to tell by a glance at it whether the current is or is not flowing in the same direction in two consecutive experiments. The galvanometer key should be a single spring key (fig. 84, page 243), not a reversing key, as that only introduces error; it should have a catch for holding the contact when required.

FRACTIONS OF A UNIT.

We have stated that the ordinary resistance boxes give R a maximum value of 10,000 units.

When fractions of a unit are required, we get one or two places of decimals by making $\frac{s}{S}$, $\frac{1}{10}$, or $\frac{1}{100}$ respectively; but this can only be done when the resistances to be measured are not more than 1000 units for one place, or 100 for two places. A further approximation can be made by observing the deflections of the galvanometer. If it is a delicate one, it will seldom be found to go truly to the zero.

Suppose that the resistance is between 1221 and 1222, then we must have $\frac{s}{S} = 1$. When R equals 1221, the spot of light will move slightly in one direction, say 5 divisions to the left; and when R is 1222, it will move in the other direction, say 10 divisions to the right.

Then we shall approximate very closely to the truth if we say that the excess of the true resistance over 1221 bears the same ratio to its defect from 1222 as the deflection to the left bears to the deflection to the right; that is, that the true resistance is 1221⅓.

DIAL-PATTERN RESISTANCE BOX.

Resistance coils are sometimes arranged as in fig. 113.

Fig. 113.

Fig. 114 shows the connections.

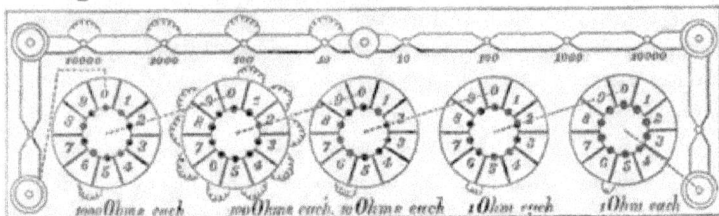

Fig 114.

The branches S, *s* are arranged in a precisely similar way to those in the ordinary boxes. The large resistance R is, however, arranged differently. Each dial consists of a brass disc surrounded by a ring cut into segments.

Each segment is connected to the next by a resistance coil, as shown in the 100 dial, fig. 114. All the coils in one dial have the same value.

The current has to pass from the segment marked 0 to the centre disc.

A plug can be placed between the centre disc and any one of the segments. If it is at the 0, there is no resistance ; but if it be placed at the 1, 2, 3 - - - - the current has to pass through 1, 2, 3 - - - coils on its way from the 0 to the centre.

The resistance in circuit is then the sum of the resistances *plugged*, and is very easily read.

Sir Wm. Thomson and Mr. Varley's Sliding Coils.

Sir Wm. Thomson and Mr. Varley have arranged a resistance box

Fig. 115.

(fig. 115) where the resistances are varied, not by inserting and taking out plugs, but by turning a handle.

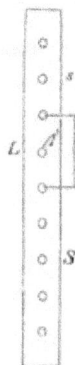

The coils are used as the branches S and *s* of a Wheatstone bridge. Any constant known resistance can be used for R.

The principle of the arrangement can be seen by fig. 116.

Fig. 117 shows the actual connections.

Let us consider fig. 116. The current from the battery is brought to a slider embracing two coils of 1000 Ohms each, and divides along the two branches, S and *s* of L. When M is moved along one division up the picture, S is increased by 1000, and *s* diminished by 1000.

Fig. 116.

To give a finer adjustment, B itself is made to consist of 100 coils of 20 Ohms each, and the battery wire can be moved along it.

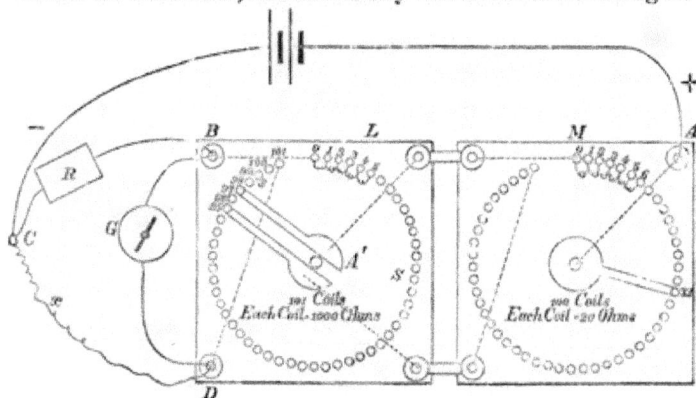

Fig. 117.

In fig. 117 we see the way in which this arrangement is practically carried out. L has two revolving arms fixed parallel to each other, and embracing two coils. The current from one pole of the battery enters this double arm at A′, passes along it, and is divided in the two branches, S and *s*. If the arm is moved one division to the right, S increases 1000 and *s* decreases 1000.

For final adjustment the current is divided in the coils M, just in the same manner as shown in fig. 116.

The resistances in the two branches, as fig. 117 is now arranged, are respectively—

$$S = 95680.$$
$$s = 4320.$$

T

RESISTANCES OF DIFFERENT SUBSTANCES.

Specific Resistance. — The specific resistance, R, of any substance is, in the C.G.S. system, the resistance between two opposite faces of a cube of the substance when the edge of the cube is one centimetre.

When we know R for each substance, we can calculate the resistance of any wire by the formula—

$$\text{Resistance} = R \frac{\text{length}}{\text{cross section}}.$$

The following tables of specific resistances of conductors and insulators are given in Professor Everett's " Units and Physical Constants," pp. 143—145 :—

As the resistances of all metals increase when they are heated, the " temperature correction " is given in each case.

Table of Specific Resistances, in Electro-magnetic Measure ∗
(at 0° C. unless otherwise stated).

	Specific resistance in absolute units.	Percentage variation per degree at 20° C.	Specific gravity.
Silver, hard-drawn . . .	1609	·377	10·50
Copper, „	1642	·388	8·95
Gold, „	2154	·365	19·27
Lead, pressed	19847	·387	11·391
Mercury, liquid	96146	·072	13·595
Gold 2, silver 1, hard or annealed . . . . . }	10988	·065	15·218
Selenium at 100° C., crystalline . . . . . . }	6×10^{13}	1·00	
Water at 22° C.	$7 \cdot 18 \times 10^{10}$	·47	
„ with ·2 per cent. H_2SO_4	4·47 „	·47	
„ „ 8·3 „ „	$3 \cdot 32 \times 10^{8}$	·653	
„ „ 20 „ „	1·44 „	·799	
„ „ 35 „ „	1·26 „	1·259	
„ „ 41 „ „	1·37 „	1·410	
Sulphate of zinc and water $ZnSO_4 + 23H_2O$ at 23° C. }	$1 \cdot 87 \times 10^{10}$		
Sulph. of copper and water $CuSO_4 + 45H_2O$ at 22° C. }	1·95 „		
Glass at 200° C.	$2 \cdot 27 \times 10^{16}$		
„ 250°	$1 \cdot 39 \times 10^{15}$		
„ 300°	$1 \cdot 48 \times 10^{14}$		
„ 400°	$7 \cdot 35 \times 10^{13}$		
Gutta-percha at 24° C. . .	$3 \cdot 53 \times 10^{23}$		
„ 0° C. . .	7×10^{24}		

∗ We remember that the Ohm (or B.A. unit) = 10^9 absolute units.

	Specific resistance.	Percentage of variation for a degree at 20° C.
Silver, annealed . . .	1521	·377
„ hard-drawn . . .	1652	
Copper, annealed . .	1615	·388
„ hard-drawn .	1652	
Gold, annealed . . .	2081	·365
„ hard-drawn . . .	2118	
Aluminium, annealed . .	2946	
Zinc, pressed	5690	·365
Platinum, annealed . .	9158	
Iron, annealed	9827	
Nickel, annealed . . .	12600	
Tin, pressed	13360	·365
Lead, pressed	19850	·387
Antimony, pressed . . .	35900	·389
Bismuth, pressed . . .	132650	·354
Mercury, liquid . . .	96190	·072
Alloy, 2 parts platinum, 1 part silver, by weight, hard or annealed . . .	2466	·031
German silver, hard or annealed	21170	·044
Alloy, 2 parts gold, 1 silver, by weight, hard or annealed	10990	·065

After many minutes electrification.	Specific resistance.	Temperature, Centigrade.	Authority.
Mica . . .	$8\cdot4\times10^{22}$	20	Ayrton and Perry.
Gutta-percha .	$4\cdot5\times10^{22}$	24	{ Standard adopted by Latimer Clark.
Shellac . .	$9\cdot0\times10^{24}$	28	Ayrton and Perry.
Hooper's material	$1\cdot5\times10^{23}$	24	Recent cable tests.
Ebonite . .	$2\cdot8\times10^{24}$	46	Ayrton and Perry.
Paraffin . .	$3\cdot4\times10^{22}$	46	Do.
Glass . . .	Not yet measured with accuracy, but greater than any of the above.		
Air . . .	Practically infinite when cold.		

The following table is due to Dr. Matthiessen :[*]—

Metal.	Percentage decrease of conducting power between 0° and 100° C.
Silver	28·44
Copper	29·69
Gold	21·30
Tin	28·89
Lead	29·61

[*] *Phil. Trans.* 1864.

T 2

GALVANOMETER SHUNTS.

For accurate determinations of resistances we must use the most delicate form of reflecting galvanometer. In cases where we do not approximately know the resistance there will, in the first few trials, be a considerable difference of potential at the ends of the galvanometer wire; and, if the corresponding current were allowed to pass, the galvanometer would be damaged. To obviate this inconvenience, " shunts " are provided.

They are also useful for measuring, by the deflection of a galvanometer, currents so strong that, if sent directly through it, they would send the spot of light off the scale.

THEORY OF SHUNTS.

The theory of shunts depends on the following deductions from Ohm's law :—

Let there be a wire a (fig. 118), and let it divide at A into two (or more) branches b and c, and let them join again at B and continue in d ; then the current in d is the same as the current in a. In the parts b and c it is divided.

Fig. 118.

Let E be the difference of potential at A and B ; let C_b C_c r_b r_c be the currents and resistances in b and c respectively ; then by Ohm's law we have in the two branches respectively,

$$E = C_b r_b \text{ and } E = C_c r_c,$$

but E has the same value for both circuits and so we have

$$C_b r_b = C_c r_c.$$

Whence we have

$$\frac{C_b}{C_c} = \frac{r_c}{r_b};$$

or, the currents in the branches are in the inverse ratio of the resistances. Thus, to construct a shunt such that the currents in the two branches shall have any given ratio, it is only necessary to make the resistances of the branches in that ratio reversed.

By connecting the poles of a galvanometer by a wire, the ratio of whose resistance to that of the galvanometer coil is known, we can send only part of the current through the galvanometer and the rest through the cross wire.

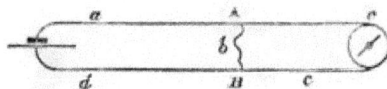

Fig. 119.

In fig. 119, the same letters are used for the same parts as in the preceding diagram, fig. 118. The reader should compare the two.

If we know the resistance of the shunt b and of the galvanometer coil c, we know from the preceding calculation the ratio of the current in the galvanometer to that in the shunt b. This, however, is not quite what we want for practical work. What we require is the ratio of the current in the galvanometer, not to that in the shunt, but to the total current of the battery—that is, to the current in a or d.

Now the whole current gets from A to B, and the only paths by which it can travel are b and c; therefore, current in b added to current in c equals whole current; or, if we call C the whole current,

$$C = C_b + C_c.$$

Let r be the total resistance of the divided circuit.

To find r, we have,

$$\text{current in } b = C_b = \frac{E}{r_b}$$

$$\text{,,} \qquad c = C_c = \frac{E}{r_c}$$

Total current,

$$C = C_b + C_c = \frac{E}{r_b} + \frac{E}{r_c};$$

reducing to a common denominator and adding, we have

$$C = \frac{E\, r_c + E\, r_b}{r_b\, r_c} = E\, \frac{r_b + r_c}{r_b\, r_c};$$

or, what is the same thing,

$$C = \frac{E}{\dfrac{r_b\, r_c}{r_b + r_c}}$$

whence

$$r = \frac{r_b\, r_c}{r_b + r_c}.$$

The current C_c in the galvanometer is

$$C_c = \frac{E}{r_c}.$$

The ratio of the latter current to the former is

$$\frac{C_c}{C} = \frac{\dfrac{E}{r_c}}{\dfrac{E}{\dfrac{r_b\, r_c}{r_b + r}}},$$

and we have from this,

$$\frac{C_s}{C} = \frac{\frac{1}{r_s}}{\frac{1}{\frac{r_s r_s}{r_s + r_s}}} = \frac{\frac{1}{1}}{\frac{r_s}{r_s + r_s}} = \frac{r_s}{r_s + r_s}$$

—that is, the ratio of the current in the galvanometer to the whole current is the same as the ratio of the resistance of the shunt to the sum of the resistances of shunt and galvanometer.

CONSTRUCTION OF SHUNTS.

In practice, shunts are usually made in sets of three, having, respectively, resistances such that $\frac{1}{10}$, $\frac{1}{100}$, or $\frac{1}{1000}$ of the whole current passes through the galvanometer, according as one or the other coil is connected to it. Of course, each galvanometer must have its own set of shunts.

Problem.—It is required to construct a set of shunts for a given galvanometer, whose resistance is known. What must be the resistance of each coil in order that $\frac{1}{10}$, $\frac{1}{100}$, or $\frac{1}{1000}$ of the whole current may pass through the galvanometer?

Case 1. Required ratio of currents 1 to 10 we have

$$\frac{C_s}{C} = \tfrac{1}{10} \qquad \text{that is}$$

$$\tfrac{1}{10} = \frac{r_s}{r_s + r_s};$$

or,

$$10 = \frac{r_s + r_s}{r_s} = \frac{r_s}{r_s} + \frac{r_s}{r_s} = 1 + \frac{r_s}{r_s}.$$

Subtracting 1 from each side, we have $9 = \frac{r_s}{r_s}$, or resistance of shunt must equal $\frac{1}{9}$ resistance of galvanometer.

Case 2. Required ratio of currents 1 to 100.

$$\frac{C_s}{C} = \frac{1}{100};$$

therefore,

$$100 = \frac{r_s + r_s}{r_s} = 1 + \frac{r_s}{r_s},$$

or

$$99 = \frac{r_s}{r_s}$$

—that is, the resistance of the shunt must be $\frac{1}{99}$ the resistance of the galvanometer.

In *Case* 3 we have required ratio 1 to 1000.

$$\frac{C_s}{C} = \frac{1}{1000};$$

therefore,

$$1000 = \frac{r_s + r_c}{r_s} = 1 + \frac{r_c}{r_s},$$

or,

$$999 = \frac{r_c}{r_s}.$$

—that is, the resistance of the shunt must equal $\frac{1}{999}$ the resistance of the galvanometer.

Galvanometer shunts are usually arranged as in fig. 120.

Three coils, having respectively $\frac{1}{9}$, $\frac{1}{99}$, $\frac{1}{999}$ of the resistance of the galvanometer, are arranged in an oblong box, from which broad strips of copper lead to the galvanometer.

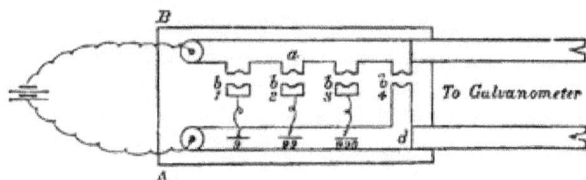

Fig. 120.

On the lid of the box A B are two blocks of brass, a and d. One end of each of the three coils is attached to d. The other ends are attached respectively to three insulated blocks, b_1 b_2 b_3. By means of a plug b_1 b_2, or b_3 can be put into connection with a.

When no plug is inserted, the whole current from the battery goes through the galvanometer. When a plug is put in at b_1, $\frac{9}{10}$ of the current flows through the first coil and only $\frac{1}{10}$ through the galvanometer. If the plug be now transferred to b_2 or b_3, $\frac{99}{100}$ or $\frac{999}{1000}$ flow through those coils respectively, leaving $\frac{1}{100}$ or $\frac{1}{1000}$ to flow through the galvanometer.

A plug placed at b_4 puts a and d into direct communication, and prevents any current passing through the galvanometer.

The same letters denote the same parts in fig. 120 as in figs. 118 and 119.

If it is necessary to leave a train of apparatus in an unlocked room, with the battery connected, the plug should always be placed at b_4, so that, in case of any one meddling with the contact keys, the galvanometer may not be injured.

In commencing to work, it is usual to put the plug at b_3, and adjust the resistance till no deflection is visible. The plug is then moved to b_2, and probably a small motion will be observed. The resistances being readjusted, the plug should be moved to b, and then taken out altogether, so that the resistance may be determined with all the accuracy of which the instrument is capable.

Fig. 121 shows a set of shunts arranged in a circular box. The copper strips are removed. The plugs are seen in b_2 and b_4.

Fig. 121.

COMPENSATING RESISTANCE.

The effect of inserting the shunt is to diminish the total resistance of the circuit, and therefore, possibly, to so increase the total current that the deflection is not reduced.

In order to keep the total current constant, a "compensating resistance" is introduced which is equal to the diminution produced by the shunt.

The introduction of the shunt reduces the resistance from

$$r_o \text{ to } \frac{r_b r_o}{r_b + r_o}.$$

The compensating resistance (ρ) must be equal to the diminution, i.e.,

$$\rho = r_o - \frac{r_b r_o}{r_b + r_o} = \frac{r_o^2}{r_b + r_o}.$$

But $r_b = r_o \dfrac{1}{n-1}$ where n is the fraction of the current which passes through the galvanometer, hence

$$\rho = \frac{r_o^2}{r_o + \dfrac{r_o}{n-1}} = r_o \frac{n-1}{n}.$$

Thus, for instance, with the $\frac{1}{10}$ shunt, the compensating resistance would have to be $\frac{9}{10}$ the resistance of the galvanometer.

SIR W. THOMSON'S METHOD OF DETERMINING THE RESISTANCE OF A GALVANOMETER.

The obvious way to determine the resistance of a galvanometer is to remove, or fix, its needle and treat it like an ordinary coil of wire, using another galvanometer in the bridge. It, however, sometimes happens that there is only one very sensitive galvanometer in a laboratory, and that it is required to know its resistance. To determine this without the necessity of procuring another instrument, the galvanometer is arranged with its needle and scale in position, in the usual position of the wire x, whose resistance is to be taken. A simple wire, whose continuity is broken by a contact key, is put where the galvanometer is usually put. In the lecture model the arrangement would be as

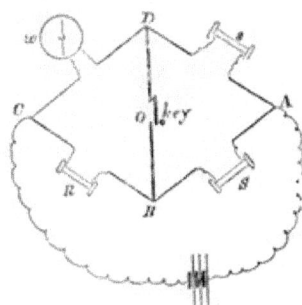

in fig. 122. Let O be the wire and key substituted for the usual galvanometer, and let x be the galvanometer whose resistance is to be measured. Let us break contact at O, and allow the current to flow in the bridge. The needle of galvanometer x will be deflected. Let us first suppose that the current in the branch A D C is so small that the de-

Fig. 122.

flection is within the range of the instrument; that is, that the spot of light does not go off the scale. Let us now observe the deflection and make contact at O. If the equilibrium is correct, no current will flow through O; if R is too great or too small, a current will flow in one or the other direction. There is no galvanometer at O, but we can tell whether or not a current is flowing in it in the following manner :—

Suppose the main current to be flowing from A to C, then, as long as contact is not made at O, the strength of the current in D C, that is in x, is equal to its strength in A D; but, if part of the current flows from D to B, the current in x will be diminished, and *the deflection will decrease* when contact is made at O.

Similarly, if a current flows from B to D, it will increase the current from D to C, and increase the deflection.

Hence the bridge equilibrium is established, that is R has its correct value, when making contact at O, neither increases nor diminishes the deflection of the needle of the galvanometer at x.

If, however, as probably will be the case, the galvanometer is so delicate that its share of the current of a single cell would send the light spot off the scale, a shunt must be applied to it. If the galvanometer has no shunts, a resistance coil, if one is at hand, or if not, a piece of wire whose resistance can be determined, may be used as a shunt.

If the resistance of the shunt wire be r_b, and that of the galvanometer r_c, we have r_b known, r_c unknown.

The result of our bridge experiment is to give us the value of the resistance of the divided circuit D to C; that is, after the experiment we know the value of $\dfrac{r_b\,r_c}{r_b + r_c}$. By substituting the value of r_b we can easily calculate r_c.

Example—

Suppose the total resistance of the divided current were 9·5, and that of the shunt wire 10.

We have

$$r_b = 10, \qquad \frac{r_b\,r_c}{r_b + r_c} = 9\text{·}5$$

Substituting the value of r_b we obtain

$$\frac{10\,r_c}{10 + r_c} = 9\text{·}5 \ \text{ or } \ \frac{10 + r_c}{10\,r_c} = \frac{1}{9\text{·}5}$$

whence

$$\frac{1}{r_c} + \frac{1}{10} = \frac{1}{9\text{·}5}$$

$$\frac{1}{r_c} = \frac{1}{9\text{·}5} - \frac{1}{10} = \frac{\text{·}5}{95}$$

and $r_c = 190$.

RESISTANCE OF BATTERIES.

The fluid of the battery itself is interposed in all battery circuits, and as it offers very considerable resistance to the current, this must be taken into account in calculating the total resistance of the circuit.

The simplest way to measure the resistance of a battery is to send its current through a tangent galvanometer with two different resistances in circuit, the resistances being chosen so as to give convenient deflections, for instance, to give say 20° in one case and 40° in the other. The ratio of the currents will then be that of the tangents of the angles of deflection. But the currents are also inversely proportional to the total resistances in circuit. From these two relations the resistance of the battery can be calculated.

For let

x be the resistance of the battery,

G   „   „   galvanometer.

a the deflection when a resistance r_a is introduced.

b   „   „ r_b „

We have

$$\frac{\tan a}{\tan b} = \frac{r_b + G + x}{r_a + G + x}$$

Let us write n for the ratio $\dfrac{\tan a}{\tan b}$, and we shall have

$$n\,r_a + n\,G + n\,x = r_b + G + x$$

or

$$(n-1)\,x = r_b - n\,r_a - (n-1)\,G$$

$$x = \frac{r_b - n\,r_a - (n-1)\,G}{n-1}$$

This method, however convenient, is not very accurate, as the resistance of the battery changes with the change of current. In the method of Mr. Mance this source of error is avoided.

Mance's Method of Determining the Resistance of a Battery.

In the arrangement described on page 280 for determining the resistance of a galvanometer, let us interchange the battery and galvanometer. It can be proved mathematically * that this will make no difference in the balancing ratios.

The same calculation as in the former case will apply, only the result gives the resistance, not of the galvanometer, but of the battery.

Mance's method was invented before Sir W. Thomson's, and the latter was derived from it.

D'Umfreville's Improvements in Mance's Method.†

The chief practical inconvenience of Mance's method is that there is always a current passing through the galvanometer, which, if the galvanometer is a sensitive one, will send the index off the scale. Although it is possible to reduce the deflection by means of magnets, the following method appears to be more convenient.

The current which in Mance's arrangement goes through the galvanometer, now goes through the wire of a coil (fig. 123)

---

* Maxwell, "Electricity," 357, vol. i. p. 413.
† "La Lumière Électrique," 1881, No. 4, p. 70.

placed inside a larger one, as explained at p. 294. The wire of
the outer coil is connected to
the galvanometer. Then however
strong a current passes through
the inner coil, no effect is produced on the galvanometer as long
as the current is constant, but at
the least variation of the current,
a current is induced in the outer
coil which deflects the needle.

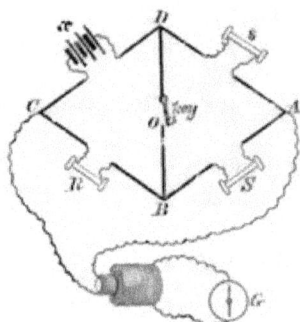

Fig. 123.

The best result is obtained
when the resistance of the outer coil equals the resistance of the
galvanometer.

Best Way to arrange Cells for any Purpose.

The fact that batteries offer resistance is the reason that large
plates are used when a strong current is required. Small plates
give as great an electro-motive force as large ones, and therefore,
by combining a sufficient number of the smallest possible cells
"in series" (fig. 124), we can get as great an electro-motive
force as we please. If, however, the cells are small, we introduce
at the same time so great a resistance, that possibly we do not
increase our current at all. In discussing the best size of cell
for any purpose, we must take into account the resistance of the
external circuit. If, as is usually the case, we have only one size
of cell, we must remember that by combining 2, 3, - - -, cells
with their like terminals connected, we produce a compound cell
equivalent to one with its plates of 2^n, 3 times - - - -, the area
of those of one cell. This method is called "connection side by
side " (fig. 125).

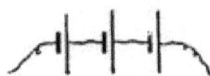

Fig. 124.
Three cells connected in series.

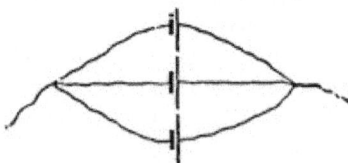

Fig. 125.
Three cells connected side by side.

If the resistance of one cell be taken as unity, the resistance of

2 cells will be	.	.	.	2	} Connected in series.	$\frac{1}{2}$	} Connected side by side.
3 „ „ „	.	.	.	3		$\frac{1}{3}$	
4 „ „ „	.	.	.	4		$\frac{1}{4}$	&c., &c.

Suppose now that we have a number of cells of a particular size, and that we wish to get the greatest possible current in a wire of given resistance.

Let E be the electro-motive force of one cell, let R be the resistance of one cell, let r be the total external resistance, and let Q be the number of cells. If the cells are arranged in series we shall have

$$C = \frac{Q\,E}{Q\,R + r}$$

If they are arranged side by side we shall have

$$C = \frac{E}{\frac{1}{Q}R + r}.$$

If they are arranged in any intermediate way, that is, in a series, each member of which consists of two or more cells arranged side by side, and we call the number of members of the series N and the number of cells in each member n, we shall have

$$C = \frac{N\,E}{\frac{N}{n}R + r}.$$

This is the general term, the two first expressions are particular cases of it.

It is obvious that Nn is the total number of cells and equals Q.

Let us now take some cases to illustrate the rule.

Let us suppose that we have 20 cells, each of 2 units resistance, and that we wish to send as great a current as possible through each of three wires separately, whose resistances are respectively 1000, $\frac{1}{4}$, and 10 units—

E depends on the nature of the cells, and is the same in all arrangements.

R = 2

r = 1000, $\frac{1}{4}$, 10 in the three cases respectively.

Case 1. r = 1000.

First arrange the whole of the 20 cells in series (fig. 126); this gives N = 20 and n = 1 ; we have

$$C = \frac{20\,E}{\frac{20}{1}2 + 1000} = \frac{20}{1040}$$

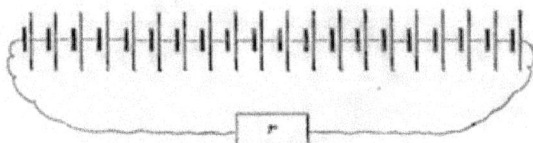

Fig. 126.

Now arrange them all side by side (fig. 127). We have

$$N = 1 \qquad n = 20;$$

and therefore

$$C = \frac{E}{\frac{1}{20} + 1000} = \frac{1}{1000 \cdot 1}$$

or only just over $\frac{1}{20}$ of what we obtained before.

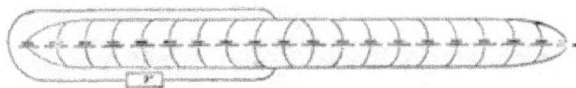

Fig. 127.

Let us now arrange the cells in 4 groups side by side, each consisting of 5 cells arranged in series (fig. 128).

This gives $N = 5$, $n = 4$, and therefore the current is

$$C = \frac{5E}{\frac{5}{4}2 + 1000} = \frac{5E}{1000 \frac{5}{8}}.$$

This is better than the last arrangement, but not so good

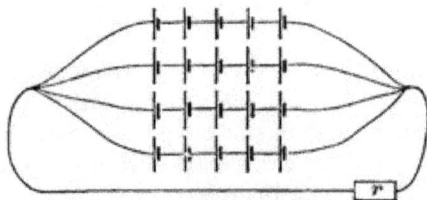

Fig. 128.

as the first; so we see that when the resistance outside is large we should arrange all our cells in series.

Case 2. Let us now try what is the best arrangement when the external resistance is small, that is when, say $r = \frac{1}{4}$.

First arrange in series—

We have

$$N = 20 \qquad n = 1 \qquad R = 2;$$

then

$$C = \frac{20\,E}{\frac{20}{1}\cdot 2 + \frac{1}{4}} = \frac{20}{40\frac{1}{4}}\,E = \tfrac{1}{2}\text{ nearly.}$$

[We see with this arrangement 3 cells give nearly as great a current as 20, for we should have

$$C = \frac{3\,E}{\frac{3}{1}2 + \frac{1}{4}} = \frac{3}{6\frac{1}{4}}\,E.]$$

Let us now arrange all the cells side by side.

We have

$$N = 1 \qquad n = 20,$$

and

$$C = \frac{E}{\frac{1}{20}\cdot 2 + \frac{1}{4}} = \frac{1}{\frac{1}{15} + \frac{1}{4}}\,E$$

$$= \frac{1\,E}{\frac{4 + 10}{40}} = \frac{40}{14}\,E = 3\,E\text{ nearly,}$$

or six times as much current as we got with the series arrangement. Therefore, when the external resistance is very small, it
is best to arrange all the cells side by side.

Case 3. When the resistance is neither very small nor very
great, we must use one of the intermediate arrangements.

Suppose $r = 10$.

First let us try series.

$$N = 20 \qquad n = 1$$

$$C = \frac{20\,E}{\frac{20}{1}\cdot 2 + 10} = \frac{20}{50}\,E = \tfrac{4}{10}\,E.$$

Now let us try "side by side."

$$N = 1 \qquad n = 20$$

$$C = \frac{E}{\frac{1}{20}\cdot 2 + 10} = \frac{1}{10\frac{1}{10}}\,E = \tfrac{1}{10}\,E\text{ nearly.}$$

Now let us try 4 sets of series with 5 cells to a series.

$$N = 5 \qquad n = 4$$

$$C = \frac{5\,E}{\frac{5}{4}\cdot 2 + 10} = \frac{5}{12\frac{1}{2}} = \tfrac{4}{10}\text{ exactly.}$$

Now let us represent the number of cells in series by the hori-

Fig. 129.

zontal distances in fig. 129 and the strength of the current by
vertical distances.

We see the strength of the current was increasing from 1 to 5, and at 20 it had the same value as at 5. It is then but reasonable to suppose that it went on increasing for a while after 5, and then began to decrease to the value it had at 20.

The point where it turned from increasing to decreasing was its maximum value.

Let us look for this intermediate point, and try an arrangement of 2 sets of 10.

We have

$$N = 10 \qquad n = 2$$

$$C = \frac{10\,E}{\frac{10}{2}\cdot 2 + 10} = \frac{10\,E}{20} = \tfrac{1}{2}\,E \text{ or } \tfrac{5}{10}\,E.$$

This then is the best practical* arrangement of cells with the given resistances.

We do not, however, in practice wish to have to do all these sums before arranging our battery ; we want a rule by which we can at once see the best arrangement.

The problem thus before us is this—

" Given R, r, and (N × n) [the number of cells] to find the ratio of $\dfrac{N}{n}$ which will make C a maximum." This problem requires the use of the differential calculus for its solution, and I have therefore worked it out in a footnote†, and can only here give the answer, which is—

---

* This is not necessarily the maximum value of the function, but it is the maximum value consistent with the experimental limitation that $\dfrac{N}{n}$ is a whole number.

† We have

(N n), R, r, E, all constant to find the value of $\dfrac{N}{n}$, which makes

$$C = \frac{N\,E}{\frac{N}{n}R + r} \text{ a maximum.}$$

This is the same as making $\dfrac{1}{C}$ a minimum—

$$\frac{1}{C} = \frac{\frac{N}{n}R}{N\,E} + \frac{r}{N\,E} \qquad \cdot \qquad \cdot \qquad \cdot \qquad (1)$$

$$= \frac{1}{n}\frac{R}{E} + \frac{1}{N}\frac{r}{E}.$$

C is a maximum when

$$r = \frac{N}{n} R,$$

that is when

$$\frac{N}{n} = \frac{r}{R}.$$

That is, to obtain a maximum current, the ratio of the number of cells in each series to the number of sets in series connected side by side, should equal the ratio of the external resistance to the resistance of each cell.

---

but $N n = $ constant, $\therefore N = \left(\frac{1}{n} . \text{constant} \right)$, and this constant $=$ number of cells $= Q$.

$$\therefore \frac{1}{C} = \frac{N}{Q} \cdot \frac{R}{E} + \frac{1}{N} \cdot \frac{r}{E} . \quad . \quad . \quad (2)$$

Differentiating (2) with respect to N we have

$$\frac{d \frac{1}{C}}{d N} = \frac{R}{QE} - \frac{1}{N^2} \cdot \frac{r}{E} \quad . \quad . \quad (3)$$

equating to zero,

$$\frac{R}{QE} \cdot \frac{E}{r} = \frac{1}{N^2}$$

or

$$\frac{R}{rQ} = \frac{1}{N^2} \quad\quad\quad (4)$$

or

$$\frac{r}{R} Q = N^2.$$

But

$$Q = N n,$$

$$\therefore N^2 = \frac{r}{R} N n$$

or

$$N = \frac{r}{R} n$$

or

$$\frac{N}{n} = \frac{r}{R} \quad . \quad . \quad . \quad (5)$$

Thus the arrangement given in the text makes $\frac{1}{C}$ either a maximum or a minimum.

Differentiating (3) again,

$$\frac{d^2 \frac{1}{C}}{d N^2} = \frac{2}{N^3} \frac{r}{E} \quad . \quad . \quad . \quad (6)$$

which is positive, after substituting for N its value obtained from (5), and remembering that N is necessarily positive; and therefore $\frac{1}{C}$ is a minimum, or the arrangement given in the text makes C a maximum.

U

This arrangement also makes *the total internal resistance equal the external,* for the equation gives $r = \dfrac{N}{n}$ R, and $\dfrac{N}{n}$ R is, as we have seen, the total resistance of the battery.

UNITS OF RESISTANCE.

In experimenting on electric resistance, it is very important to have some unit of resistance to which other resistances can be referred. The first and obvious form of unit is a wire of some fixed length, section, density, and substance at a given temperature.

The Siemens unit consists of mercury at a temperature of $0°$ C. enclosed in a glass tube 1 metre long and 1 square millimetre in section.

This is a purely arbitrary standard, and cannot be connected with any absolute system of measurement. A system of measurement based on the C.G.S. system has been introduced by the British Association, and the unit derived from it is called the B.A. unit or Ohm.

We must defer the consideration of this to a later portion of the book, as to understand it we require a knowledge of some portions of electrical science which we have not yet treated of. (See Chapter XXVII., vol. i. p. 303.)

CHAPTER XXIII.

Equivalent Magnet.

If a current of electricity be sent round a very small ring of wire, the latter acts in all respects as a short magnet would do if it had been suspended at the centre of the coil, and had taken up its natural position when acted on only by the current. That position we remember is perpendicular to the plane of the ring, and with its marked end to the left of a man supposed to be swimming round the ring, down stream looking towards the centre.*

If, keeping the strength of the current constant, we make it circulate round two or more concentric rings, we increase the magnetic moment of the equivalent magnet by increasing its strength. If the rings are arranged in a helix, we increase the arm of the couple and so increase the moment.

Electro-Magnet.†

If a bar of soft iron be placed in the axis of a coil of wire, and a current sent round the coil, the bar becomes a magnet with its marked end to the left of a man swimming down the current looking towards the axis of the coil.

When the current ceases, the bar loses its magnetic properties; when it is reversed, the magnetism of the bar is reversed. The magnetism of the bar increases when the strength of the current is increased. At first, when the current is feeble, it increases approximately at the same rate. As the strength of the current increases, the strength of the magnet increases in a ratio, smaller and rapidly diminishing, until at length a point is reached where increasing the current produces no further increase of magnetism

---

* See p. 241.
† See Chapter XXXVI.

in the bar.   When this occurs, the bar is said to be "saturated with magnetism."

MUTUAL ACTION OF TWO CURRENTS.

Two currents in the same direction attract each other.

Two currents in opposite directions repel each other.

The force between two parallel straight currents is numerically equal to the product of the strength of the currents multiplied by their length, divided by the square of the distance between them.

This fact was discovered by Ampère.

ELECTRO-MAGNETIC INDUCTION.

The following discovery is due to Faraday :—

MOTION OF WIRE.

If a wire be moved in the neighbourhood of a magnet in any direction, except along a line of force, a difference of potential will be produced at the ends, which, if the ends be connected by a wire not acted on by the inducing magnet, will cause a current.*

The direction in which the current will flow may be remembered as follows :—

If in the northern hemisphere a person with arms extended moves forward, then the current which would tend to be produced, in a wire represented by his arms, by the action of the earth's magnetism would flow from his right hand to his left.†

MOTION OF MAGNET.

If a magnetic pole be moved in the neighbourhood of a wire in any direction except parallel to it, a current will be induced

---

* There are several ways in which this can be managed. The most obvious is to let the ends of the moving wire slide on two fixed rails (fig. 130) connected by a fixed cross wire—

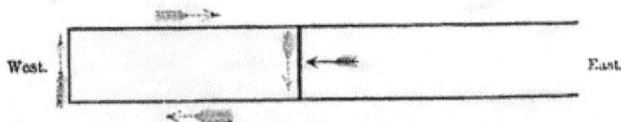

West.   East.

Fig. 130.

⇒——→ Motion of wire.
⇒⇒----→ Current due to a marked magnetic pole in front, or under the influence of terrestrial magnetism.

† Faraday, Exp. Res. 3079, vol. iii. p. 332.

in the wire. If, for instance, a magnet S N, fig. 131, be lifted
suddenly in and out of a coil of wire, a current will be induced

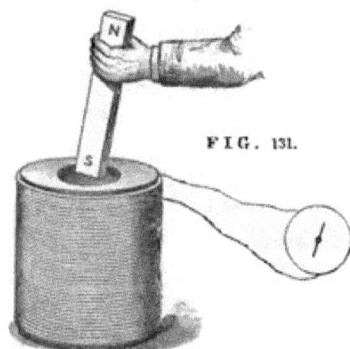

FIG. 131.

which will be in one direction on inserting the pole, and in
the other on drawing it out. If the magnet be reversed so
as to use the other pole, the current will be reversed.

MOTION OF CURRENT.

If, instead of a steel magnet, a coil of wire carrying a
current, as in fig. 132, be lifted in or out of the outer coil,

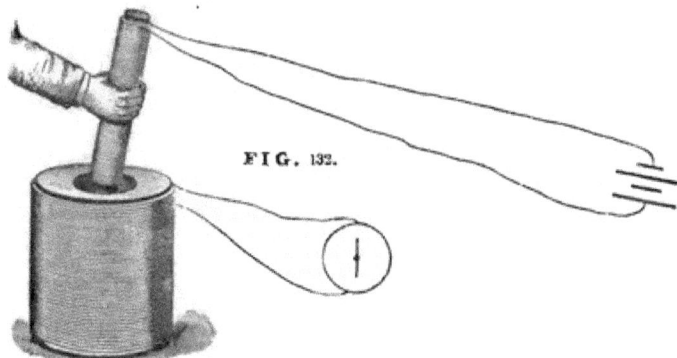

FIG. 132.

currents will be induced whose directions depend on the directions
of motion of the inner coil and on the direction of current in
it. The inner coil acts in all respects as its equivalent magnet
would do.

In both cases the current is stronger the more suddenly the
magnet or coil is lifted in and out, and in the case of the
moving wire (p. 292) it is stronger as the wire moves faster.

VARIATION OF CURRENT.

If, in fig. 132, the smaller coil is left inside the larger, and the current in it made to vary, then at every increase a current will be induced in one direction in the outer coil, and at every decrease an equal current will be induced in the other direction, for varying a stationary current produces the same magnetic effect as moving a steady current.*

If a core of metal be placed inside the inner coil, the effect on the outer one will be increased.† If the core is of iron, the effect will be very much increased.

* No effect whatever will be produced in the larger coil, however strong the current in the smaller is, as long as it remains constant.

† See p. 298.

CHAPTER XXIV.

THE TELEPHONE AND MICROPHONE.

BELL'S TELEPHONE.

THE now well-known Bell Telephone is a beautiful application
of the theory of electro-magnetic induction. Fig. 133 shows it in
section, fig. 134 in perspective. It consists of a steel magnet S N,

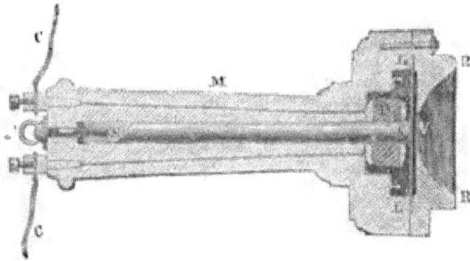

Fig. 123.

round one end of which is wound a coil of fine wire B. The
magnet and coil are enclosed in a wooden tube M, one end of
which, R V R, is of considerably greater diameter than the magnet.

Across the wide end of the tube a diaphragm of thin sheet
iron L L is fixed, which just does not touch the pole of the
magnet.

When the instrument is spoken to, the iron plate vibrates in
time with the sound vibrations.

As it moves it causes temporary alterations in the magnetism
of the steel magnet, and these in turn induce periodic currents
in the coil of wire.

The induced currents are conveyed along the telegraph line
C C, and received in a similar telephone at its other end.

They travel round the coil of wire in it, and cause temporary changes in the magnetism of the steel magnet.

Owing to these changes, the force with which the iron plate is attracted varies, and the latter is caused to vibrate in time with the vibrations of the plate of the sending instrument.

The plate, as it vibrates, sets the air in motion and reproduces exactly not only the note but the words spoken into the sending instrument. The voices of different speakers can be recognized even at a distance of many miles. Descriptions of various modifications of the instrument will be found in works on telegraphy.*

When a *steady* battery current is sent through a telephone, no sound is produced, but every *variation* of the current causes a loud noise. The instrument is sometimes used to determine whether a given current is constant or intermittent.

Fig. 134.

THE HUGHES MICROPHONE.

When at any point in a circuit carrying a battery current there is an imperfect contact, any change in the goodness of the contact will produce a change in the current and cause a sound in a telephone included in the circuit.

Professor Hughes has discovered that, when the imperfect contact consists of two pieces of carbon lightly pressed together,

* See Du Moncel, *Le Téléphone, le Microphone, et le Phonographe* (Hachette, Paris, 1878).

variations in the current are caused by the very smallest sound occurring near the carbon.

The *microphone* (fig. 135) consists of two or more pieces of car-

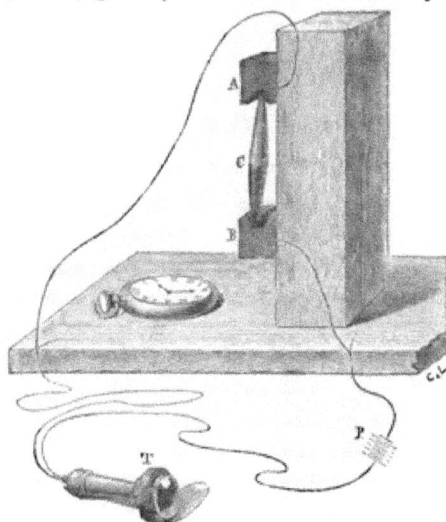

Fig. 135.

bon A C B lightly pressed together. A telephone T and a battery P are included in circuit with it.

The lowest whisper spoken near the microphone is loudly reproduced in the·telephone.

To intensify the effects, the microphone is usually placed on a sounding-board. The sound caused by a fly walking on the sounding-board is distinctly audible at the telephone.

The ticking of a watch sounds like blows of a hammer.

CHAPTER XXV.

HUGHES' VOLTAIC INDUCTION BALANCE.——PLATE XXV.

IN the spring of 1879 Professor Hughes communicated to the Physical Society* an account of a *Voltaic Induction Balance*, invented by him for the purpose of measuring the conductivity for instantaneous induced currents possessed by various substances, and " for the investigation of the molecular construction of metals and alloys."

When two coils of wire are placed near together, and a pulsating current is sent through the one, a certain current will be induced in the other every time that the current in the first coil alters.

If a metal core be placed inside the coils, the induced current will produce an increased sound in a telephone in circuit, and, in general, the amount of increase will be greater as the quantity of metal and its conductivity increases.

In the induction balance, Plate XXV., there are two primary oils, $a\,a'$, and two secondary, $b\,b'$.

The same pulsating current is sent through the two coils, $a\,a'$, and the secondary coils are so connected to each other, that the induced currents in them are in opposite directions, and, when equal, exactly neutralize each other.

The primary current in $a\,a'$ is produced by three Daniell cells, and its pulsations are caused by a microphone inserted in the circuit. The sounds which excite the microphone are produced by the ticking of a clock.

The equality of the induced currents is tested by means of a telephone inserted in the joint circuit of b and b'.

When the induced currents are equal, the telephone is silent, but the least inequality in them causes it to sound.

If two exactly equal pieces of the same metal, such as two new shillings, are placed inside the coils $a\,b$, $a'\,b'$ respectively, no effect will be produced; but if there is the least difference be-

* *Phil. Mag.*, July, 1879, ii. page 50.

Sliding Coil

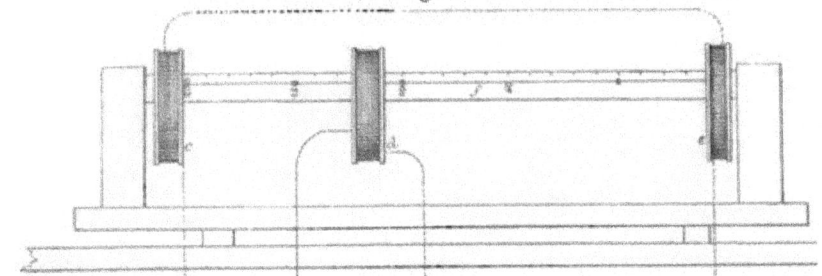

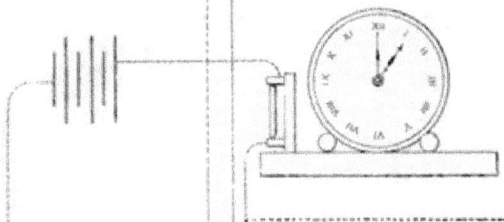

INDUCTION BALANCE

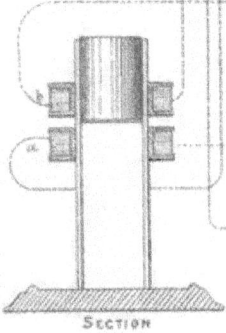

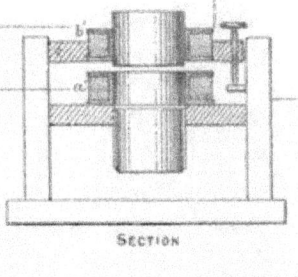

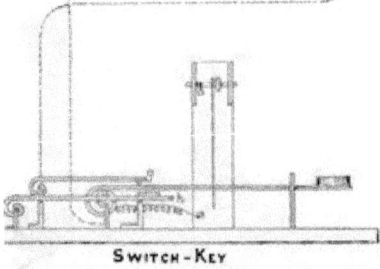

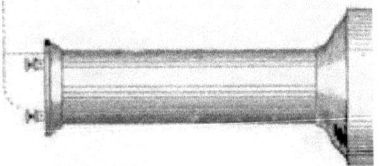

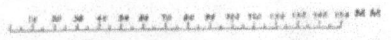

SECTION

SECTION

SWITCH-KEY

Plate XXV. Hughes' Voltaic Induction Balance.

tween them—if, for instance, one of the shillings is a little worn
—the telephone will sound loudly.

In order to measure the differences of induction produced by
different pieces of metal, the " sliding coil"[*] shown at the top of
Plate XXV. is used.

At each end of a divided bar f, coils c and e are fixed, round
which the primary current travels in opposite directions. A
third coil, d, can be slid along the bar, and is in circuit with b,
one of the secondary coils.

When d is exactly midway between c and e, no effect is produced on it, as their actions on it balance; but when it is slid
nearer to one or the other, a current is induced in it which either
increases or diminishes the current in b according to the direction
in which d has been moved.

The distance which d has to be moved from the centre to
establish the balance, when equal and similar pieces of different
metals are placed in $a\,b$ and $a'\,b'$ respectively, gives in arbitrary
units of the instrument the difference between the instantaneous
conductivities of those metals.

The instrument is so sensitive that the insertion of "a milli-
gramme of copper or a fine iron wire, finer than a human hair,"
on one of the coils causes the telephone to sound loudly.

The inventor states that he has already, by the aid of the instrument, studied "the effects on metals, of heat, magnetism,
electricity, &c., and of mechanical changes such as strain, torsion,
and pressure," and he proposes " in some future paper to describe
the remarkable results already obtained."

Mr. Chandler Roberts has shown[†] that by means of this instrument it is possible to test the fineness of alloys.

A silver-gold alloy, containing only two grains of gold to the
pound Troy of silver, can be clearly distinguished from pure
silver by means of the balance.

The instrument can be used to detect bad coins, as, if a good
sovereign be placed in one coil and a bad one in the other, the
telephone instantly sounds.

The tones produced in the telephone are found to be different
when different metals are used to disturb the equilibrium; thus iron
gives "a dull smothered tone, hard steel an exceedingly sharp one."

[*] Called by Prof. Hughes a " Sonometer."

[†] *Phil. Mag.*, July, 1879, ii. page 57,

PROFESSOR HUGHES' RESEARCHES ON MOLECULAR MAGNETISM.

APPARATUS.

On continuing researches with the induction balance, Professor Hughes found that though it was always easy to obtain a balance when two similar pieces of non-magnetic metal were placed in the two coils respectively, yet that when two pieces of iron or steel were used, it was almost impossible to obtain the balance. This led him early in 1881 * to examine whether there was any special peculiarity in these metals which would account for the difference observed.

For this new investigation it was found necessary to modify the balance as follows :—

A single coil of wire is used, and the straight wire under examination is stretched so as to pass through its centre. The coil is movable on a pivot, so that the wire can either lie exactly on its axis or be inclined at a considerable angle to it. The intermittent current can be sent through the wire and a telephone connected to the coil, or the current can be sent through the coil and the telephone connected to the wire. This last was the arrangement generally used.

The exterior diameter of the coil is 5½ centimetres, and its width 2 centimetres. The opening in the centre of the coil is 3½ centimetres in diameter. The coil is wound with 200 metres of No. 32 wire.

One end of the straight wire is fixed, and the other is attached to a torsion head, so that the wire can be twisted as desired. We see that if the straight wire is of copper or any non-magnetic metal, the current in the coil would have no effect on it if it lay exactly in the axis of the coil, for then the straight wire and the wires of the coil would be exactly at right angles to each other. If the coil were inclined so that its axis no longer coincided with the straight wire, then the intermittent current in the coil would induce currents in the wire which would get stronger as the inclination increased, i. e. as the wire and the wires of the coil became more nearly parallel. If, however, the straight wire were wound into a spiral, and placed with its axis parallel to the axis of the coil, currents would be induced in it by the intermittent current in the coil; and further, if

* Proc. Roy. Soc., vol. xxxi., 1880-81, p. 525.

without the whole wire being twisted into a spiral, its molecules
were so re-arranged that a current passing from end to end
would have to travel spirally round it instead of in a straight
line along it, we should have induced currents.

If, then, we wish to determine whether a straight wire has
such a spiral structure or not, we can do so by placing it in the
axis of the coil and seeing if currents are induced.

Effect of Elastic Twist.

When an iron wire was used, it was found that any elastic
twist produced sounds in the telephone, i. e. showed either that
currents were travelling spirally round the wire, or that its
molecules were so arranged as to be the "equivalent magnets" *
of such currents. This effect was only produced by the elastic
twist, for when a permanent twist was given to the wire, the
effect produced was extremely small; in fact, the effect was in
general proportional, not to the twist, but to the amount of
untwisting which took place under the influence of elasticity
when the wire was released. No such effect was produced by
copper or any other non-magnetic wire. Thus we see that in
the iron or steel wire an elastic twist produces a spiral structure
in the wire which causes a current to pass from end to end of it,
to circulate round it, and magnetize it.

Spiral Structure produced by a Current.

Professor Hughes found that if a current was passed through
an iron wire, it produced a spiral structure causing currents to
be induced in the coil, and that this structure was permanent.
It was exactly similar in its effects to that previously produced
by an elastic twist, and, indeed, could be removed by mechanically
untwisting the wire; the current, however, did not produce any
visible mechanical torsion, i. e. on one end of the wire being
released from its clamp after the passage of the current, it did
not show any tendency to rotate.

* Vol. i. p. 291

CHAPTER XXVI.

MEASUREMENT OF ELECTRO-MAGNETIC INDUCTION.

WE stated in Chapter XXIII. that, "if a wire be moved in the neighbourhood of a magnetic pole, in any direction except along a line of force, a difference of potential will be produced at its ends; which difference of potential can, under certain circumstances, produce a current."

We know that the electro-motive force will be increased if we move the wire more rapidly or increase the intensity of magnetization.

Let us suppose that the lines of magnetic force, in the region through which the wire is to move, have been drawn, and let us express the intensity of the magnetization, at any surface perpendicular to the lines of force, by the number of them, which pass through a square centimetre of the surface.

The fact that the intensity of magnetization of any surface was equal to unity would thus be expressed by drawing one line of force through each square centimetre of it.

We now see that if a straight wire be moved, without revolution, uniformly through the field, the difference of potential produced at its ends will bear a constant ratio to the number of lines of force cut in a second.

For the number of lines in each square centimetre will be increased if we increase the magnetic force; and if we increase the velocity of the wire, the number of centimetres passed over in a second will be increased. By suitably choosing our units (as has been done in the C.G.S. system), we can make this constant ratio one of equality; and we may state the following proposition :—

If a straight wire be moved uniformly in a magnetic field, a difference of potential will be produced at its ends, which will be numerically and algebraically equal to the number of lines of force cut in a second, it being remembered that lines of force cut in one direction are counted (+), *and the same lines, if cut by a motion in the opposite direction, are counted* (—).

If the wire be not straight, the difference of potential at its ends will be the same as that at the ends of a straight wire whose ends coincide with the ends of the bent wire.

Corollary.—No current will be produced in a wire forming a closed circuit which moves parallel to itself, if the field is uniform over the whole extent of the circuit.

If the wire be not moving uniformly, but with variable velocity, the difference of potential at its ends at any instant will be equal to the number of lines which it would have cut if it had gone on moving uniformly for one second with the velocity that it had at that instant in a field whose intensity continued equal to that of the actual field at the instant under consideration.

Certain cases of rotatory motion will be considered later on.

If the wire be moving in a direction perpendicular to the lines of force, it will cut more of them in moving a given distance, than if it moved in any direction more nearly parallel to them ; and if it is moving along them, it will cut none of them, and no difference of potential will be produced.*

If the ends of the wire be connected by another wire, which is not in motion, as, for instance, if the wire slides on two fixed rails connected at one end, a current will flow through the system.

* The ratio of the differences of potential produced when the wire is moving in any plane, to that produced when it is moving in a plane perpendicular to the lines of force, is expressed by the cosine of the angle between the planes.

CHAPTER XXVII.

BRITISH ASSOCIATION UNIT OF RESISTANCE CALLED THE "OHM."

IN experimenting on electric resistance, it is very important to have some unit of resistance to which other resistances can be referred. The first and obvious form of unit is a wire of some fixed length, section, density, and substance at a given temperature.

The Siemens unit consists of mercury at a temperature of $0°$ C. enclosed in a glass tube 1 metre long and 1 square millimetre in section.

This is a purely arbitrary standard, and cannot be connected with any absolute system of measurement. In order to obtain a material standard whose value is known to be a convenient decimal multiple of the absolute electro-magnetic unit of resistance, the system of measurement which we are about to explain has been introduced by the British Association.

We have stated a relation between the motion of a wire in the electro-magnetic field, and the electro-motive force between its ends.

The current in the wire is, as we know, the electro-motive force divided by the resistance.

The absolute electro-magnetic unit of resistance is the resistance which satisfies Ohm's law when the electro-motive force is equal to the current, both being expressed in electro-magnetic measure; for we see that in the equation

$$E = C\,r,$$

making $C = E$ involves the condition $r = 1$.

If a wire whose ends are connected by a wire at rest and without resistance be moving in a magnetic field, so that the number of lines of force cut by it in a second is equal to the current produced in it, the resistance of that wire must be unity. If the current be measured by the deflection of a needle in the same magnetic field, it becomes unnecessary to know the intensity of magnetization in the determination of the equality of the potential and the current, as we shall show that this, being a factor on both sides of the equation, is cancelled out. In actual determinations, the earth's magnetism supplies a field whose intensity is uniform over a sufficiently large region.

RAILS AND SLIDER.

Let us now return to our system of rails (page 292). The method we are about to describe cannot be carried out, but it is introduced for illustration, and a modification of it is used experimentally.

Let us consider the fixed connecting bar to be curved into an arc of a circle (fig. 136), and let the rails be one centim. apart, and be in the same vertical plane.

Fig. 136.

To simplify calculation, let this plane be magnetic E. and W.

Let now a straight vertical wire, whose resistance can be varied, move along the rails at the rate of one centim. per second, viz. with unit velocity.

Let the arc and the rails be supposed to have no resistance.

At the centre of the arc, which, for simplicity, we will suppose to be a half circle, whose plane is at right angles to the plane of the rails—that is, vertical and magnetic N. and S.—let a small needle be suspended. The arc and needle will then form a tangent galvanometer, only, as it has only half a ring, its equation will be

$$C = H \tan \delta \, \frac{a}{\pi} \quad \cdots \cdots \quad (1)$$

or as $2\,a = 1$ centim.

$$C = H \tan \delta \, \frac{1}{2\,\pi} \quad \cdots \cdots \quad (2)$$

As the length of the slider is unity, the number of lines it cuts per second will be equal to its velocity multiplied by the earth's horizontal magnetic force.

When the slider moves, a difference of potential is then produced at its ends which is equal to

Earth's Horizontal force × velocity of slider.

The only resistance in circuit is the resistance of the slider. The strength of the current produced is then, by Ohm's law, equal to

$$C = \frac{\text{Earth's Horizontal force} \times \text{velocity of slider}}{\text{Resistance of slider}}.$$

This tends to deflect the needle with a certain force. The earth's magnetic force pulls it back, and the moments of these two opposing couples vary with the angle of deflection. When the needle is at rest, the couples are equal.

When the needle is at rest at 45°, tan $\delta = 1$, and the equation (2) becomes

$$C = \frac{1}{2\,\pi} H \quad \cdots \cdots \quad (3)$$

Hence we have,—

When the needle is at 45°,—

Earth's Horizontal force $= 2\,\pi$ current.

That is

x

$$\left.\begin{array}{l}\text{Earth's Horizontal}\\\text{force}\end{array}\right\} = 2\,\pi\ \frac{\text{Earth's Horizontal force} \times \text{velocity of slider.}}{\text{Resistance of slider}}$$

Divide both sides of this equation by earth's horizontal force and we have

$$1 = 2\,\pi\ \frac{\text{velocity of slider}}{\text{Resistance of slider}}.$$

Multiply both sides by resistance of slider and we have

$$\text{Resistance of slider} = 2\,\pi\ \text{velocity of slider.}$$

We see that this result is independent of the intensity of the earth's force.

The unit of resistance is then the resistance of that wire which, if used as a slider to connect the two rails above mentioned, would allow a velocity of the slider of $\dfrac{1}{2\,\pi}$ *centims. per second to deflect a needle at the centre of the half ring 45°.*

This is the absolute electro-magnetic unit. It is, however, of inconveniently small dimensions; and therefore the B.A. unit or "Ohm" is defined as being equal to 10^9 (a thousand million) absolute electro-magnetic units.

A mile of pure copper wire No. 16 gauge has a resistance of about 13.7 B.A.U.

PRACTICAL METHODS.

We will now examine the methods by which the unit is practically determined.

It is obvious that the arrangement with the rails and slider cannot be carried out. We must examine in what way we can modify it, so as to make it experimentally possible.

The first necessity is to substitute circular for rectilinear motion, so as to keep the apparatus in one place during the operation. The next is to dispense with the rails and connecting bar. We must remember that if our slider had a resistance of only one B.A. unit, we should, to get a deflection of 45°, require a velocity of $\dfrac{1}{2\,\pi}$ thousand million centims., or nearly 1000 miles per second.

REVOLVING SEMICIRCLE.

If we place a semicircular wire, pivoted at A and B (fig. 137), with its diameter vertical and its plane coinciding with an equipotential surface of horizontal magnetic force, i. e. magnetic east and west, and cause it to turn half round on its diameter in one second, so that at the end of the second it is in the position shown by the dotted lines, it will have described a hemisphere, and the projection of this hemisphere on a plane perpendicular to the lines of force (the plane of the paper) will be the circle ACBC'.

The number of lines of force cut during the first second is then the number which pass through this circle.

If now the motion is continued in the same direction of revolution, the other hemisphere will be described in the second second. The number of lines of

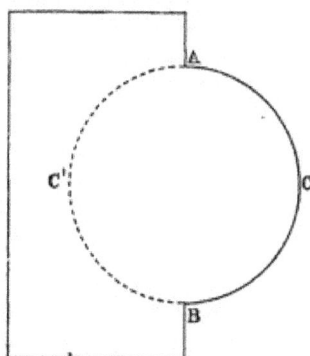

Fig. 137. Elevation.

force cut will be the same as before; but as, during the first second, the wire was moving from E. to W., and during the second from W. to E. (fig. 138), the electro-motive force will,

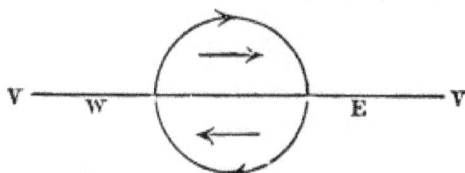

Fig. 138. Plan.

though numerically the same, be algebraically opposite; and if the first velocity be called +, the second will be —.

Thus, in the first half of the revolution, a current is induced downwards* in the wire, while, in the second half, an equal current is induced in the opposite direction— that is, upwards. By suspending a small needle at its centre, we cause this revolving ring to become its own galvanometer. At the same time, as the current reverses, the position of the wire with regard to a magnet sus-

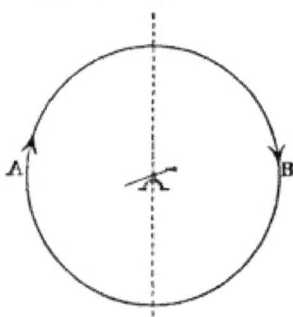

Fig. 130

* See vol. i. p. 292.

x 2

pended at its centre reverses also, so the current will always
tend to deflect the magnet in the same direction. ·

For, let A B (fig. 139) be a vertical section of an ordinary
tangent galvanometer, and let us consider separately two semi-
circles separated by a vertical diameter. The current at A is then
upwards, that at B downwards; and they both tend to deflect the
needle in the same direction, and the two halves of the fixed ring
are in precisely the same state as the revolving semicircle in its
two positions.

REVOLVING RING.

We have hitherto supposed the current of the semicircle to
complete its circuit by way of the supports. This, however, is
not necessary; for if we make the revolving wire a complete
circle, each half completes the circuit for the other, and we get
double the effect on the needle. In order to still further increase
the effect on the needle, the simple ring is, in actual measurements, replaced by a coil of a large and known number of turns.
Such was the apparatus used in the determination of the absolute unit of resistance by the Committee of the British Association appointed to report on Electrical Standards in 1861.

REPORT OF THE COMMITTEE.

The following members of the Committee, appointed by the
British Association, presented, in 1863, a description of an
"Experimental Measurement of Electrical Resistance made at
King's College, London,"[*] from which the following pages are
extracted :—

Members of the Committee :—Professor J. Clerk Maxwell,
Messrs. Balfour Stuart and Fleming Jenkin.

The description of the apparatus which is quoted is by Mr.
Fleming Jenkin.

The experiments were made in June, 1863. We must premise that when the ring revolves with uniform velocity, the deflection is perfectly steady; for though the force exercised on the
needle is different at different points of revolution, yet, the variations being periodic, and having a very small period (for the
velocity of revolution is great), the force on the needle is always
sensibly equal to the mean force.

[*] *Reports on Electrical Standards*, p. 97. Edited by Prof. Fleming
Jenkin, F.R.S. (Spon, 1873).

The second portion of the Report describes the methods of constructing resistance coils equal to the revolving coil. The following is the

" Description of the Apparatus:"—
Plate XXVI.

" For convenience of description, the apparatus with which the experiments were made may be divided into five parts :—(1), the driving gear ; (2), the revolving coil ; (3), the governor ; (4), the scale, with its telescope, by which the deflections of the magnet were observed ; (5), the electric balance by which the resistance of the copper coil was compared with a German-silver arbitrary standard.

" The general arrangement of the first four parts is shown in Plate XXVI.

The Driving Gear.

" *The driving gear* (fig. 4) consisted of a leaden fly-wheel X, on a shaft A turned by hand, and communicating its motion by a band, $b\, b_1 b_2 \ldots$, arranged in a way equivalent to Huyghens' gearing, to a shaft B, a pulley on which drove the revolving coil by a simple band $a\, a_1 a_2$. The arrangement of the band $b\, b_1 b_2 \ldots$ communicating the motion of the shaft A to the shaft B may be easily understood from the diagram. C C are two guide pulleys running loose on pins attached to the main framing. D D are two loose pulleys maintained at a constant distance by the strut E, to which the weight W is hung.

" When the rotation of the shaft B is opposed by a sufficient resistance, the effect of turning the fly-wheel in the direction shown by the arrow is to lift the weight W from the ground, tending to turn the shaft B with a definite force, which will be sensibly constant so long as the weight is kept off the ground, and the band $b\, b_1 b_2 \ldots$ is kept unaltered in length. Wherever, as in the present experiments, the resistance increases with the speed of rotation, the speed of the driving-wheel can easily be regulated by hand, so as to keep the weight from falling so low as to touch the ground, or rising so high as to foul the gear ; and thus, with a little care, a constant driving force can be applied to the shaft B, and to the machinery connected with it."

THE REVOLVING COIL.

"The revolving coil* formed the most important part of the apparatus. It is shown one-fifth full size in figs. 1 and 2, Plate XXVI.

"A strong brass frame HH was bolted down by three brass bolts, F F F, dowelled into a heavy stone. It could be accurately levelled by three stout screws, G G G.

"The brass rings, I I, on which the insulated copper wire was coiled, were supported on the frame by a pivot, J, working in lignum vitæ, and by a hollow bearing, K (fig. 1), working in brass : this bearing worked in a kind of stuffing-box, k (figs. 2, 3), which, by three screws and a flat spring washer between it and the frame at J, could be adjusted to fit the collar e with great nicety, preventing all tendency to bind or shake. Supported in this way, the coil revolved with the utmost freedom and steadiness.

"The coil of copper wire was necessarily divided into two parts on the two rings, I I, to permit the suspension of the magnet S.

"The two brass rings were each formed of two distinct halves, insulated from one another by vulcanite at the flanges *ff*. This insulation was necessary to prevent the induction of currents in the brass rings.

"These rings, after being bolted together, were turned with great accuracy by Messrs. Elliott Brothers. The insulated copper wire was wound in one direction on both rings; the inner end of the second was soldered to the outer end of the first; the two extreme ends of the conductor thus formed were soldered to two copper terminals, *hh'*, insulated by a vulcanite piece, *x*, bolted to the brass rings. Each terminal was provided with a strong copper-binding screw, and had a mercury-cup drilled into its upper surface. The two coils could be joined so as to form a closed circuit, by a short copper bar between the binding screws. The bars, binding screws, and nuts were amalgamated to ensure perfect contact. When the copper coils were to be connected with the electric balance, the short copper bar was removed, and the required connections were made by short copper rods $\frac{1}{4}$ inch in diameter, dipping at one end into the mercury-cup on the terminals, *hh'*, and at the other end into the mercury-cups of the electric balance.

* Now deposited in the Cavendish Laboratory, Cambridge.

"The absence of all induced currents influencing the suspended magnet, when the circuit was broken at hh', was repeatedly proved by experiment.

"Rotation was communicated to the coils by a catgut band, simply making half a turn round the small V-pulley l. The band could be tightened as required by the jockey pulley z and weight w (fig. 4).

"A second V-pulley, r, served for the band cc, communicating motion to the governor by which the speed was controlled."

THE COUNTER.

"A short screw n, of large diameter, gearing into a spur-wheel of 100 teeth, o, formed the counter from which the speed of rotation was obtained, as follows :—A pin, p, on the wheel, o, lifted the spring, q, as it passed ; this spring, in its rebound, struck the gong, M. The blow was of course repeated at every 100 revolutions, and the time T of each blow was observed on a chronometer. The arrangement was equally adapted for rotation in either direction.

THE SUSPENDED MAGNET.

"The manner in which the suspended magnet was introduced to the centre of the coil is best seen in fig. 3. A brass tripod N, bolted to the main frame, supported the long brass tube O, which passed freely through the hollow bearing at k. A cylindrical wooden box, P, slipped on to the lower end of the tube O.

"The magnet hung inside this box, the lower part of which could be removed, to allow the exact position of the magnet to be verified. The support N also carried a short brass tube R, on which the glass case T could be secured by a little sliding tube. The mirror t, attached to the magnet S by a rigid brass wire, hung inside this glass case by a single cocoon fibre about seven feet long. This fibre was protected against currents of air by a wooden case (not shown in the plate), extending from the point of support down to the glass case. A little sliding paper prolongation of the wooden case made it nearly wind-proof, by fitting at the bottom against the main brass frame. An opening in the case allowed the mirror to be seen. The fibre at the top was suspended from a torsion-head, by which it could be turned ; it could also be raised and lowered by a small barrel, and was adjustable in a horizontal plane by three set screws. The care taken

in suspending the magnet, and in protecting it both against currents of air and vibration, was repaid by success, for the image of
the scale reflected in the magnet was as clear and steady when
the coil was making 400 revolutions per minute as when it was
at rest. The governor used was lent by one of the committee, and
generally controlled the speed to such uniformity as allowed the
deflections to be observed with as much accuracy as the zero
point."

THE SCALE AND TELESCOPE.

"The scale* and telescope hardly require special description—
they were arranged in the usual manner for this kind of experiment at about 3 metres from the mirror. The scale was an
engine-divided paper scale nailed to a wooden bar. This plan
will in future experiments be abandoned, as variations in the
weather had a very perceptible influence on the scale."

THE ELECTRIC BALANCE.

"The annexed diagram (fig. 140) shows the electric balance,
by which the resistance of the revolving coil II, Plate XXVI.,
R, fig. 140, was compared with that of an arbitrary German silver

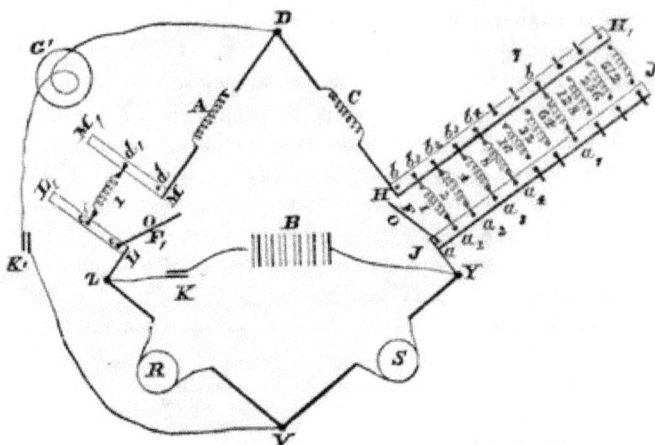

Fig. 140.

standard S, before and after each rotation experiment. The
arrangement is a modification of the ordinary Wheatstone's

* See vol. i. p. 169.

balance.* A and C represent the branches of the balance, S the German silver standard, and R the copper coil to be measured.† JJ_1, HH_1, MM_1, and LL_1, are four stout copper bars with mercury cups a a_1 a_2—b b_1 b_2—c c' and d d'. Two short copper rods, F and F_1, can be used to connect a with b, and c with d. When this is done, the arrangement is exactly that of the Wheatstone balance, p. 248, with keys at K and K'. A and C were coils formed of about 300 inches of No. 31‡ German silver wire, and were adjusted to equality with extreme nicety, and each assumed equal to 100 arbitrary units.

" If R on any occasion had been exactly equal to S, the galvanometer G' would have been unaffected on depressing the keys K K' when a was joined to b and c to d by F and F', rods of no sensible resistance.

" This exact equality between R and S could never be obtained, owing to slight changes in temperature which affected the two coils very differently. The object of the modifications introduced was to allow the ratio between S and R differing by a small amount only, to be measured with great accuracy.

" For this purpose a number of German silver coils were adjusted representing $1.2.4.8 \ldots 512$ in the arbitrary units equal to the hundredth part of A or C. These coils were so arranged that any one of them could be introduced between the bars HH_1 and JJ_1.

" A single coil, equal to 1 in the same arbitrary unit, could be introduced between the bars LL_1 and MM_1. In this diagram this coil is shown in its position, and the rod F' withdrawn. Similarly F is withdrawn from between H and J', and the coil 1 joins a and b in the bars HH_1 and JJ_1. If no other coils were placed between HH_1 and JJ_1 the arms of the balance would now be 101 and 101 respectively instead of 100 and 100; but the ratio would still be that of equality. Let us now suppose that, when the circuit with the battery is completed, the galvanometer

---

* Vol. i. p. 262.

† To compare with fig. 101 we have

In fig. 140.		In fig. 101.
A	corresponds to	S
C	"	s
R	"	R
S	"	z

‡ Diameter = 0·01 inch.

by its deflection shows that R is bigger than S, we can reduce the resistance of the arm between D and Y by various small graduated and definite amounts, by introducing the coils 2, 4, 8, &c., between HH_1 and JJ_1. Let us first suppose the coil 2 introduced. The resistance between H and J will be the reciprocal of 1.5, or 0.6667; for where various resistances are added in multiple arc, the resistance * of the compound arc is the reciprocal of the sum of their conducting powers, and the conducting power of a wire is the reciprocal of its resistance. The ratio between the two arms will now be 101 : 100.6667. Let us suppose that on completing the circuit the galvanometer still deflects in the same direction as before, the arm between D and Y must be still further reduced by including fresh coils between HH_1 and JJ_1. It is very easy by trial to find the combination which maintains the galvanometer at zero when the circuit is completed. Let us suppose that, as in the diagram, the coils included were 1, 2, 4, 8, and 64. The reciprocals of these numbers are 1, 0.5, 0.25, 0.125, and 0.015625. The conducting power between H and J is therefore 1.890625, the sum of these numbers. The resistance between H and J is 0.52893, the reciprocal of the last number, and the ratio between the arms will be 101 : 100.52893. A little consideration will show that with the coils named any ratio between 101 to 100.5 and 101 to 101 can be obtained by steps not exceeding 0.00195, the reciprocal of 512, the largest coil or smallest conducting power which can be included between the copper bars HH_1 and JJ_1. By substituting the rod F_1 for the coil 1 between LL_1 and MM_1, the observer can obtain a fresh series of ratios with the same steps between 101 to 100 and 100.5 to 100. In this way it will be seen that unless the coils R and S differ by more than one per cent., their ratio can be measured in the manner described within 0.002 per cent.

" It should further be observed that extreme accuracy in the coils 1, 2, 4, &c., is not necessary, since an error of one per cent. in the sum of these, as compared with their true relative value to the coil C, would only affect the final result 0.01 per cent.

" The position of R and S in the balance relatively to A and C, &c., is, of course, interchangeable.

" The diagram is not intended at all to represent the practical arrangement, but simply to show the connections. The coils

* Vol. i. p. 276.

1, 2, 4, &c., had amalgamated copper terminals, which simply dropped into mercury cups on the copper bars; the observations could be made very rapidly and accurately, as the galvanometer was sensitive-enough with four Daniell's cells to indicate the addition or subtraction of the 512 coil with perfect distinctness.

" The reduction of the observations to find the ratio seems somewhat complicated at first, but with the aid of a table of reciprocals it takes but little time. No improvement seems necessary in this part of the apparatus. The idea of using large coils combined with small ones in multiple arc, to obtain extremely ·minute differences of resistance, was suggested by Professor [now Sir] W. Thomson, and will be found useful in very many ways."

THE CONTACT KEY.

" In all bridge experiments it is necessary to make contact in the battery circuit, before making it in the galvanometer circuit, and thus to avoid the extra current * produced immediately on making contact. It is also important that the battery current should not flow longer than necessary, as by heating the coils it alters their resistance. The key here shown (fig. 141)

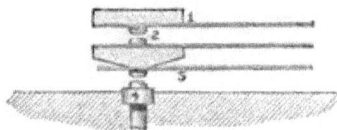

Fig. 141.

is contrived to make the battery contact just before the galvanometer contact, and to break it just after. The upper springs, 1, 2, give the battery contact, the lower, 3, 4, that of the galvanometer. The drawing now explains itself.†

DETAILS.

" The following are some of the details of the experiments made at King's College, June, 1863 :‡—

n, the number of windings was 307 ;

l, the effective length of wire 302·063 metres;

D, the distance from the mirror to the scale 2·9853 metres.

---

* See vol. i. p. 329.

† For the mathematical theory of the experiments, see *Reports on Electrical Standards*, p. 101.

‡ *Reports on Electrical Standards*, p. 104.

"DETERMINATION OF DEVIATION.

"δ is the difference between the reading of the scale when the magnet is acted on by the earth only, and when it is acted on also by the induced currents in the coil. To determine δ the reading of the scale is made when the coil is at rest, or when the circuit is broken. Another reading is taken with the connection complete and the coil in motion. If the *direction* of the earth's magnetism remains the same, the difference of these readings is the true value of δ; but since the direction of the earth's magnetic action is continually varying, we must find the difference of *declination* between the times of the true readings, and calculate what would have been the undisturbed reading of the scale at the time when the deviation was observed. In these experiments this correction was made by comparison with the photographic registers of magnetic declination made at Kew, at the same time that the coil experiments were going on." *

CORRECTIONS.

Corrections were made for—A, the dimensions of the sections of the coil ; B, for level.

In the actual experiments the level was taken with a spirit-level, reading to 12″, and found correct to at least that degree of accuracy.

C. Correction for the induction of the suspended magnet on the coil. The strength of the magnet, as compared with that of the magnetic field, was measured by means of a magnetometer from Kew by the ordinary method.†

D. Correction for torsion of fibre.‡ "This correction depends on the relation between the stiffness of the fibre and the directive force of the suspended magnet. The fibre was a single fibre of silk, 7 feet long; the magnet was a steel sphere ₁₆ inch diameter, and not magnetized to saturation. The correction for torsion was therefore much larger than if a stronger magnet had been used."§

" E. Correction for position of suspended magnet."

F. Correction for irregularity in the magnetic field due to iron or magnets near the instrument.

---

* See vol. i. p. 189.
† Ibid., p. 168.
‡ Ibid., p. 174.
§ *Reports on Electrical Standards*, p. 105.

It was found that this correction was so small that it could be neglected.

G. Correction for scale reading.

The scale is supposed to be divided into millims; if, however, it stretched or shrank, it would no longer be accurate. The error being measured, correction G is applied.

H. Correction for electro-magnetic capacity of coil.

FURTHER DETAILS.

The account of experiments continues :[*]—

" The nature of the electrical action in the experiment may be stated as follows :—

" Suppose the plane of the coil to coincide with the magnetic north and south, and that the coil is revolving in the direction of the hands of a watch. Then the north side of the coil is moving from west to east, and therefore experiences an electro-motive force tending to produce an *upward* current.[†] The south side of the coil is moving from east to west, and therefore there is a tendency to produce a downward current in it. If the circuit is closed, there will be a current upwards on the north side and downwards on the south side round the coil.

" Now, this current will tend to turn the north end of the suspended magnet towards the east; but the earth's magnetic force tends to turn it towards the north; so that the actual position assumed by the magnet must depend on the relation between the strength of the current and the strength of the earth's magnetism. But the strength of the current depends only on the velocity of rotation, the resistance of the coil, and the strength of the earth's magnetism. Hence the position of the magnet will not depend on the strength of the earth's magnetism, but only on the velocity and the resistance of the coil.

" We must remember that the coil in its revolution comes into other positions than that which we have mentioned. As the north side moves towards the east, the current continually diminishes till it ceases when it is due east. The current then commences in the opposite direction with respect to the coil; but since the coil itself is now in a reversed position, the effect of the current on the suspended magnet is still to turn the north

---

* *Reports on Electrical Standards*, p. 106.

† Compare vol. i. p. 306, " Revolving semi-circle."

end to the east. The **action** of the current on the **magnet is, therefore,** of an intermittent **nature,** and the **position of the** magnet is **not** fixed, but **continually** oscillating. The extent of these oscillations, however, is exceedingly **small.**

"The whole **extent** of the vibration would be less than $\frac{1}{100}$ of a millim. on the **scale.**

"This vibration **was never observed,** and did not interfere with the distinctness **of** vision.

"The **only** oscillations observed were **the** free oscillations of the **magnet.** They arose from accidental causes at the beginning of the experiment, and **were subject to slight** alterations in **magnitude due to speed of rotation, the passage of iron steamers in the Thames, &c.**

"The **time** of one vibration was about **9.6** seconds, and by reading the scale at the **extremities of every vibration a** series of **readings** was obtained. **The intervals between each** were approximately **equal.** All we **have to do is to observe the** deviation at **every** oscillation, and **to ascertain the whole** number of revolutions **during** the time of observation, and the exact beginning **and ending of that time.** This was done in the following way :—

METHOD OF WORKING.

"The coil **was made to revolve by means of the** driving **machine, and its** velocity was regulated **by the governor.** While **the required velocity** was being attained, **the oscillations of the magnet were reduced within convenient limits by means of a** quieting **bar at a distance. The quieting bar was then put in** its **proper place, and the observation commenced.†**

"One **observer, A, took the readings of the scale as seen in the telescope, writing down the deviation at the extremity of every** oscillation, **and thus obtaining a reading every 9.6 seconds.**

"Another **observer, B, with a chronometer, wrote down the times of** every third **stroke of the bell.** The times thus found were **at intervals of 300 revolutions. When the observer B** noted **the** time, **the observer A made a mark on his paper, so that after the** experiment **the** readings **of deviation could be compared with the readings of the** chronometer taken **at the same** time.

---

* These experiments were made before the construction of the Embankment, when King's College was close to the water.

† Vol. i. p. 180.

" The mean time of revolution between any two times of observation could thus be found and compared with the mean deviation between the same limits of time, and any portion of an experiment accidentally vitiated could be rejected by itself.

" The experiments of each day commenced with a comparison, by means of an electric balance, between the resistance of the experimental (revolving) coil and that of a German silver coil (called Time 4). Then a series of readings of the scale was taken, to determine the undisturbed position of the magnet. The times of beginning and ending this series were noted, and called times of 1st Zero.

" Then the coil was made to revolve, and readings of deviation and of time were taken, as already described, and called 1st Spin +.

" Then the direction of rotation was reversed, and a second set of readings obtained, and called 2nd Spin —.

" Then the undisturbed position was again observed, with a note of the time. This last was called 2nd Zero.

" Lastly, the resistance was compared again with the standard coil. This series of experiments was then repeated, if there was time.

" From the values of 1st and 2nd Zero, together with the information obtained from the photographic resisters at Kew, the true value of the undisturbed reading during the 1st and 2nd spin was obtained. The difference between this and the actual reading is the deviation δ, due to the electric currents.

RESULT.

From these experiments, by a mathematical calculation, the resistances of the arbitrary German silver coils were determined in absolute measure. The absolute resistance of any one of them being known, coils of given resistances can be constructed by the ordinary methods. The absolute resistance of the German silver coil marked " June 4, 101 " [arbitrary units] was found to be 107620116 metres per second.

ISSUE OF B.A. UNITS.

This determination being completed, the Committee proceeded to construct material copies, each having a resistance equal to one thousand million centimetres per second, that is—

$$10^9 \text{ C.G.S. units.}$$

This larger unit is called sometimes the "British Association unit," written "B.A.U.;" sometimes the "Ohm," after the discoverer of the law of electrical resistance.

METALS AND ALLOYS SELECTED.

An important question arose as to what was the best material of which to manufacture the coils.

The following is the answer to it by the Committee,* taken from Appendix A, by Dr. A. Matthiessen, F.R.S., and Mr. Ch. Hockin, to the report of 1865 :—

"Several unit-coils have been made and issued.

"We propose to state the method by which these coils were made, and the reasons for choosing the particular alloy which has been adopted for the conductor.

"The alloy referred to is composed of 66 per cent. of silver and 33 of platinum.

"This alloy possesses many properties which fit it for the use to which it has been put.

"As to its electrical properties :—

"I. It alters less in electrical resistance with changes of temperature than any other known alloy.

"The importance of this point needs hardly to be enforced on any one who has used resistance coils.

"The increment in the resistance of the alloy, due to a change of temperature from 0° to 100° C., is only 3.2 per cent.

"II. The conducting power of the alloy is very low, and is about one-half that of German silver.

"III. The conducting power of the alloy is not altered by baking, i. e. by exposing it to a temperature of about 100° C. for several days.

"This is a property of great importance, for it has been observed that those conductors which do not alter by baking do not by age either. The experiments by which this has been established have been published in former reports.

"IV. The conducting power of a wire of the alloy is little altered by annealing. Further, the alloy does not oxidize by exposure to the air; it does not readily alloy with mercury; it makes a sufficiently pliable wire, and can be drawn to a very great degree of fineness. Of this alloy twenty unit-coils have been

* *Reports on Electrical Standards*, p. 135.

made and sent to several leading electricians at home and abroad. The form of bobbin adopted for putting up the wire (shown in Plate XXIV.) has been found very convenient, as it can be immersed in water during an observation. The wire is twice coated with silk, and protected by being imbedded in solid paraffin.

"Besides the coils already mentioned, ten unit-coils have been made, which will be deposited at the Kew Observatory.

"Any one possessing a copy of the B.A.U. may have it compared at any future time against one of these coils for a small payment.

"Of the coils to be sent to Kew two are of the platinum-silver alloy, two of a platinum-iridium alloy, and two of commercially pure platinum.

"Two mercury units have also been prepared.

"With so many coils for reference, made of such different metals, it appears quite improbable that the unit now proposed should be lost.

"Along with the above-mentioned coils will be preserved the standard coil used in the experiments first referred to, the coil used in the similar experiments made by the Committee in 1863, and several copies of these coils.

"Of the coil called 'June 4,' in the Report for 1863, two German-silver copies have been made. Of the other coil used in 1864 two German silver, two gold-silver, and one platinum-silver copy have been made."

STANDARD COMPLETED AND DEPOSITED AT KEW.

"These coils have twice been re-compared together at intervals of three months, and will be again compared; and if they are still found not to have altered they will be deposited at the Kew Observatory for reference, their values being engraved on them.

All the coils to be issued are re-compared some weeks after they are made, and rejected if they are found to have altered in resistance 0·01 per cent.

"All the coils sent out are correct at the temperature written on them to *within* 0·01 per cent., and this temperature lies between 14·5° and 16·5° C. in all cases."

For ordinary work the units are made of German silver, as it is nearly equal to the platinum alloy and very much cheaper.

Redetermination by Lord Rayleigh and Dr. A. Schuster.

On April 12, 1881, Lord Rayleigh and Dr. A. Schuster communicated to the Royal Society[*] an account of a re-determination of the value of the Ohm made by them in the Cavendish laboratory. They used the same apparatus as that used by the British Association Committee, but it was re-adjusted and improved. They also carefully examined the various formula of correction used in the first determination, and in particular the correction H. (p. 317) for the electro-magnetic capacity of the coil; that is, for what is called the "Self-Induction."[†] They found that in the first experiments this correction had been made too small. The effect of this would be to give a value of the Ohm less than its true value.

The authors give as the final result of their experiments that the British Association Ohm is more than 1 per cent. too small, and that its value is

$$\text{B.A. Ohm} = \cdot 9893 \text{ True Ohm,}$$

or that

$$\text{True Ohm} = 1\cdot 0108 \text{ B.A. Ohm.}$$

On February 15, 1882, Lord Rayleigh communicated his final results to the Royal Society.[‡] The apparatus used was of the same kind as that used by the British Association Committee, but larger and more accurately made. The value obtained is

$$1 \text{ B.A. Ohm} = \cdot 98651 \text{ True Ohm.}$$

or

$$\text{True Ohm} = 1\cdot 01367 \text{ B.A. Ohm.}$$

Rowland's Redetermination.

Professor Rowland has also redetermined the Ohm. His result is

$$\text{B.A. Ohm} = \cdot 9926 \text{ True Ohm,}$$

or

$$\text{True Ohm} = 1\cdot 0074 \text{ B.A. Ohm.}$$

Mercury Standards.

The following are the lengths of a column of mercury at $0°$ C and of one square millim section which represent the different

---

[*] Proc. Roy. Soc., vol. xxxii., 1880-81, p. 104.
[†] See page 317.
[‡] Phil. Trans. 1882, p. 661.

determinations of the Ohm. The specific resistance of mercury is given at 96,190 by Everett,* and at 96,146 by Maxwell.† We have then—

Mercury column at 0° C, and one square millim section.	Length in centimetres.	
	Maxwell.	Everett.
B.A. Ohm	104·08	103·96
Lord Rayleigh's { Determination of April 12, 1881	105·20	105·08
Ohm. { Determination of Feb. 15, 1882	105·51	105·37
Rowland's **Ohm**	104·86	104·73
Paris Congress Ohm		

The last line of this table cannot be filled in until the report of the International Commission is published. Whatever length they give to the standard mercury column is to be definitely adopted for all practical work, and is not again to be re-corrected, but is to be the legal and commercial standard in future. Similarly, as we have already explained, if new determinations of the earth's quadrant show that the metre is not exactly its ten-millionth part, the metre will not be altered, but the earth's quadrant will be said to consist of more or less than ten million metres.

* "Units and Physical **Constants**," p. 144.
† "Electricity," § 362, vol. ii. p. 418.

CHAPTER XXVIII.

MUTUAL ACTION OF CURRENTS ON EACH OTHER AND BETWEEN CURRENTS AND MAGNETS.

WE have stated that if two wires carrying currents be freely suspended, a force will be observed between them which will be attractive if the currents are in the same direction—repulsive if they are in opposite directions.

Wires being suspended so that they can turn and move without much friction, and without stopping the battery currents, the actions are observed.

Experiment has shown that the action of a small plane circuit is the same, at distances which are great, in comparison with the diameter of the circuit, as that of a small magnet inside it having its pole in the position that, if previously magnetized, it would take up under the influence of the current, and having a magnetic moment equal to the product of the strength of the current into its small area.

Thus, any very small plane circuit may be considered as equivalent to a magnetic shell bounded by the wire, and whose magnetic intensity is equal to the strength of the current. The marked side of the shell is on the side to which the marked end of a magnet placed inside would turn.

The following demonstration is due to Professor Maxwell :*—

The magnetic action of any closed circuit whatever is the same as that of a magnetic shell bounded by the wire, and of a strength equal to the strength of the current.

First, suppose the area bounded by the wire to be entirely filled up with very small circular currents (fig. 142)—that is, small rings put as close together as possible, then smaller ones in the spaces

---

* *Electricity*, 483, vol. ii. p. 131.

left between them, then smaller ones in the residual gaps, and so on till the whole surface is covered; and let a current of the same intensity, and in the same direction as the main current run round each; then each is equivalent to a magnetic shell of its own area and of the same strength as the current; and all these shells added together make one shell of the same strength as the current, and whose area is equal to that of the original circuit.

But the magnetic effect of this system of circuits is equal to that of the original circuit, for " the magnetic effect of two equal and opposite currents flowing close together is absolutely zero." Now every current in these rings, except those in the outside portion of the outside rings, has another equal and opposite current flowing close to it, and therefore the magnetic effect of the whole system becomes that of a current of the same strength as the original current, and whose shape can, by increasing the number and diminishing the size of the imaginary rings, be made to differ from that of the original current by as small a quantity as we please.

Fig. 142.

And, as things which are equal to the same thing are equal to one another, we have shown that any current forming a closed circuit is equivalent to a magnetic shell of the same strength, and whose edges coincide with the wire.

We stated in Part II.* that the magnetic action of a magnetic shell on any point outside it depends only on its strength and the solid angle subtended by its bounding edge, and therefore not on the shape of the shell; and we now see that this is the case if we consider the equivalent current; for the small circuits which are equal to the shell, and also to the original current, will continue to be equal to the latter in whatever way they are drawn, whether on a plane surface or one curved in any way. The only requirement is that each is close to the one next to it—a condition which can be fulfilled equally well by a surface of any form.

To calculate the magnetic potential at any point due to a current forming a closed circuit—

Substitute for the circuit its equivalent magnetic shell. We then calculate as for a magnetic shell.

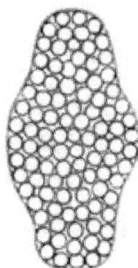

* Vol. i. p. 158.

CHAPTER XXIX.

RELATION BETWEEN VARIATION OF POTENTIAL AND STRENGTH OF CURRENT.

Ohm's law tells us that, when the resistance is constant, the current in any wire varies directly as the difference of potential at its ends.

Let us draw a horizontal line AB, figs. 143, 144, 145, whose

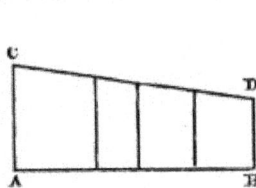

Fig. 143.

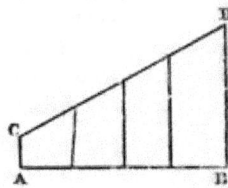

Fig. 144.

length represents the resistance of a wire carrying a current; and let us draw vertical lines at various points whose lengths

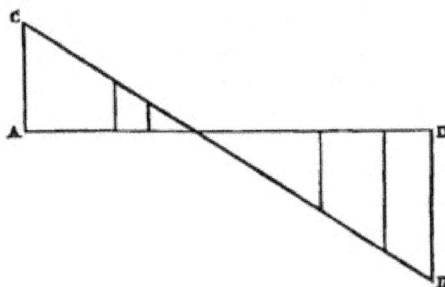

Fig. 145.

represent the potentials at those points—positive potentials being drawn upward, negative ones downward.

By Ohm's law the extremities of these ordinates will all lie in the same straight line CD. If the difference of the potentials at the ends is great, the slope of the line CD will be steep; if it is small, the line will be more nearly horizontal. That is to say, the slope of the line CD represents the *rate of variation of potential* as we travel along the wire from A to B.

But as the line CD is straight, its slope depends directly on the difference of height of its ends, and inversely on the length of the line AB.

By Ohm's law the strength of the current varies directly with the difference of potential at the ends of the wire, and inversely with its resistance. Therefore the slope of the line CD represents the strength of the current in the wire.

But as we have just shown, this slope also represents the rate of variation of the potential. Therefore, the strength of the current in AB is proportional to the rate of variation of the potential in it; and as AB may be made as small as we please, we may say that the current *at* any point in a wire is proportional to the rate of change of the potential as we begin to leave that point.

Finally, by so choosing our units, as we have done on the C.G.S. system, we may make the ratio one of equality, and say—

*The current at any point in a conductor is equal to the rate at which the potential varies as we begin to leave that point.**

Of course the direction of the current is given by the direction of the slope of the line CD.

---

* That is—If V be the potential at the point, the current is $\frac{dV}{ds}$ where s is a length measured along the wire.

CHAPTER XXX.

CURRENTS PRODUCED BY INDUCTION ON CLOSING AND BREAKING THE CIRCUIT.

Relative Motion.

The experiments described on pages 292 and 293 have shown us that if we move a wire in the neighbourhood of a magnet, a certain difference of potential is produced at the ends of the wire.

We also saw that the same effect will be produced if we leave the wire stationary and move the magnet, as there is the same relative motion in each case.

Instead of a magnet of ordinary form, we may use a magnetic shell; and, finally, we may replace the magnetic shell by its equivalent current, as in fig. 142, p. 325, and in fig. 132, p. 293.

Now we have shown that the potential at any point due to a circuit is equal to the solid angle subtended at that point multiplied by the strength of the current; so that, if we vary *either* the solid angle or the strength of the current, we vary the potential.

When we move the wire, we have shown that we cause an electro-motive force in a wire in its neighbourhood as long as the motion continues; and similarly, if we vary the current, we cause a similar electro-motive force as long as the variation continues.

Variation of Current.

Increasing the current has the same effect as approaching the wire; diminishing the current has the same effect as moving the wire away.

Gradual Increase.

When the ends of a wire are connected to a battery, the current which flows through the wire does not at once attain its maximum value, but gradually increases for a short time (a small fraction of a second). Similarly, when the current is interrupted, the diminution is somewhat gradual.

In each of these cases a current will be induced in any neighbouring wire, which current will be in opposite directions on making and breaking contact, and will be nearly instantaneous in duration. The direction of the induced current, on making contact, is the same as that which it would have if the inducing current, flowing constantly in the same direction, were brought nearer to the inducing wire ; that is, it is in the opposite direction to the primary current.

We see that the strength of the induced current will be influenced by the rate at which the potential of the secondary wire is made to vary, because that is proportional to either the rate of variation of the solid angle, or the strength of the inducing current.

EXTRA CURRENT.

If a current be sent through a *coil* of wire, it is observed that, on closing the circuit, the current does not instantly reach its maximum value, and that, on breaking the circuit, it does not instantly fall to zero. This effect is the same as would be produced if, at the moment of closing, a transient current were produced in the wire in the opposite direction to the primary, and, at the moment of opening, another in the same direction as the primary.

The same effect would also be produced if the phenomena of the electric current were due to the motions of a fluid having inertia, and therefore momentum when in motion. If this were the case, the fluid would, when the electro-motive force was applied, begin to move slowly by virtue of its inertia, and, when the electro-motive force ceased, its momentum would prevent the resistance from at once stopping its motion.

Many experiments have shown that the first of these hypotheses is the correct one ; or, at least, that the first hypothesis is the true explanation of by far the greater portion of the effect. There is still, however, a minute residual effect about which experiment has not yet given a decided answer.

The transient currents in a coil are produced by the induction of each portion of the current on the neighbouring wires, on which it acts as if they were portions of another circuit.

These transient currents are called the "_Extra Currents_" of closing and opening respectively.

If the coil be unwound and stretched out (fig. 146), so that no part is near any other part, except at right angles to it, the extra currents almost entirely disappear. This is because, for two wires to act inductively on each other, it is necessary that they should be near together, and not at right angles. I say _almost_ the whole extra current disappears (that is, the main current attains its maximum or its zero _almost_ immediately after closing or opening), because there is a slight residual effect while the wire is in what was called by Faraday the _electro-tonic_ state—that is, the changing state while the current is forming or ceasing.

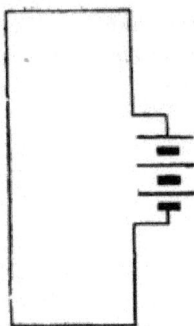

Fig. 146.

MEASUREMENT OF CURRENTS PRODUCED ON CLOSING AND OPENING THE CIRCUIT.

Except in certain special cases, the strength of the induced current at any instant is not measured, owing to its short duration, and, seemingly, almost arbitrary variations, even during the short time it lasts.

In measuring the induced current, both these difficulties are avoided, by measuring, instead of its value at one instant, the sum of all the values it has during the short time it lasts.

This is managed by using a galvanometer with a somewhat heavy needle, which, on receiving a sudden impulse, begins to move very slowly. When the induced current passes each value, it gives a proportional impulse to the galvanometer needle, and the last impulse has been received by it before the first has sensibly displaced it from its position of rest.[*]

The impulses, being all added together, cause the needle to swing from its position of rest. The extreme limit of the swing is a measure of the sum of the impulses—that is, of the total current produced by closing or opening a neighbouring circuit.[†]

[*] The balistic galvanometer (vol. i. p. 255) is used for experiments of this nature.

[†] If _i_ be the strength of the induced current at any instant, the total

BLASERNA'S EXPERIMENTS.

Professor Blaserna [*] has investigated the laws of currents of opening and closing, and the duration of the currents produced, both on opening and closing, both in the wire itself, and in other wires in the neighbourhood.

His method of proceeding was as follows :—

To measure the duration of a current in the secondary wire, it is necessary to disconnect the secondary wire from the galvanometer at a small known interval of time after the closing or

SCALE OF INCHES

Fig. 147.

opening of the primary circuit. The shortest value of this interval which is found not to affect the strength of the secondary current is the duration of that secondary current.

THE DIFFERENTIAL INTERRUPTOR.

Professor Blaserna makes this measurement by means of two cylinders, C C′ fig. 147, clamped on a common axis, the surface of

current is
$$\int_o^t i\, dt$$

where t is a time not less than that which the primary current takes to vary from zero to its maximum value on closing, or from its maximum value to zero on opening the circuit.

[*] Sullo sviluppo e la durata delle correnti d'induzione, e della estracorrenti Professor Pietro Blaserna. *Giornale di Scienze Naturali ed Economiche.* Vol. vi., 1870, Palermo.

part of each being wood, part metal. Two metal springs press on the two cylinders, one on each, and respectively close or open the primary and secondary circuits, according to whether they press on wood or metal in each case. By causing the cylinders to revolve rapidly, and with known velocity, any duration of contact, however short, can be obtained at will.

Fig. 147 gives the details of the apparatus, which the inventor calls the " Differential Interruptor."

Fig. 148 is a diagram of the same apparatus.

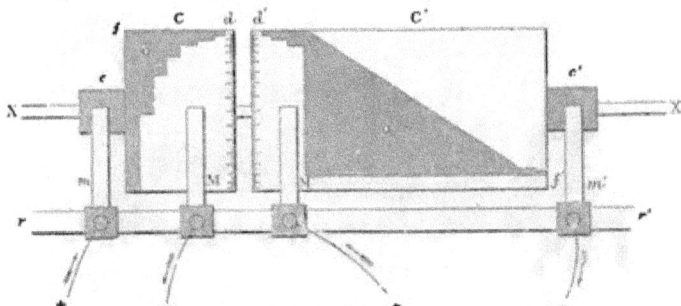

Fig. 148.

Round the pulley-wheel A passes a band from a heavy fly-wheel, which is turned by hand by an assistant. By means of a train of cog-wheels, R*r* R′*r*′, a very rapid motion is communicated to the cylinder. M M′ are the contact springs, which can be moved longitudinally, and clamped in any position. The cylinders can be placed in any position with regard to each other, the position being known by the divided circles *dd*′.

The springs *m m*′ which press on the rollers *c c*′ complete the circuits.

The metal portion of the surface of the cylinder is, on the primary cylinder, cut into steps; on the secondary, partly into steps and partly on a slope.

By suitably placing the springs M M′, we can cause the secondary circuit to be closed or opened when the cylinder has turned through any desired angle, after either the closing or the opening of the primary circuit.

MEASUREMENT OF THE VELOCITY OF REVOLUTION.

A wheel B is fixed to the axis of the cylinder, and turns with it. In its circumference are several rings of holes, the holes in each ring being equidistant from each other. The rings contain 96, 48, and 24 holes respectively.

A jet of air from an india-rubber tube plays upon the wheel, and is alternately checked or allowed to pass freely, according as a hole, or solid metal, comes opposite to the mouth of the jet. When these openings and closings succeed each other rapidly, a musical note is produced.

Now the number of vibrations per second of a tuning-fork or other instrument producing a note is known. When, therefore, the note produced by the air-jet is the same as that of a tuning-fork which gives n vibrations per second, we know that n holes pass the aperture in a second, and therefore that the number of turns per second is

$$\frac{n}{96}, \frac{n}{48}, \text{ or } \frac{n}{24},$$

according to which row of holes we are using.

An observer, with a sufficiently good musical ear, can determine the velocity more accurately by this than by any other method.

Now suppose the cylinder turning with a velocity of N turns per second; and that between, say—closing the primary current and opening the secondary, we have to turn the cylinder θ degrees, then the time which elapses between closing the primary and opening the secondary will be

$$\frac{\theta}{360} \cdot \frac{1}{N} \text{ of a second,}$$

which, if we are using, say the outside row of holes, will be

$$\frac{\theta}{360} \cdot \frac{96}{n} = \frac{4}{15} \cdot \frac{\theta}{n} \text{ of a second,}$$

where n is the number of vibrations per second of the note given by the perforated wheel.

PRELIMINARY EXPERIMENTS.

Professor Blaserna first describes a number of preliminary experiments to show that the instrument may be trusted to make contact exactly at the times when the edges of the brass come under the springs, and not a fraction of a second later. This point being satisfactorily established, he goes on to the experiments which form the real object of his research.

The room in which the experiments were made was about 18

feet square, and the "Differential Interruptor" was bolted to a
stone slab fixed to one wall of the room, while the galvanometer
stood on another similar slab at the other side of the room, and
was therefore not affected by the vibration of the machinery.

One of the galvanometers was a tangent galvanometer furnished with a mirror. A method of graduating it, that is of
eliminating its errors and determining exactly what strength of
current corresponds to each deflection, is given. An astatic galvanometer is also used in the observations.

The first experiments were made on the

INDUCED CURRENT OF CLOSING.

In these, the arrangement was that of fig. 149. The primary
current was closed for 180°, and the secondary current was

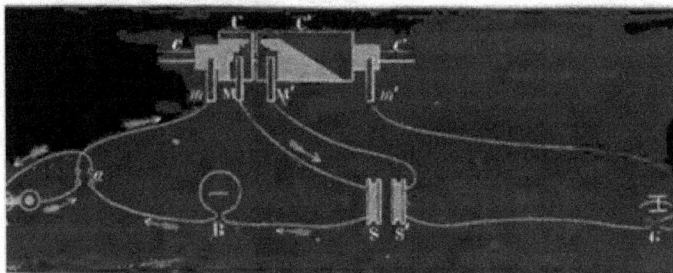

Fig. 149.

broken when the cylinder had turned 1·9° after closing the
primary.

It was found that when the circuit was closed, and a current
passed which caused a deflection of 25° in a tangent galvanometer
B, that the impulse due to the closing deflected the second galvanometer G, $3\frac{1}{2}°$. The coils SS' were distant 1 centim. from each
other. On turning the wheel gently so that the cylinder made
3·76 revolutions per second, the deflection of the induction
galvanometer at once went up to 36°, showing that the effects
on the needle of the quickly recurring impulses were added together.

As the velocity increased the deflection increased, until at
9·07 turns the deflection was 42°. This was its maximum, and
it then began to decrease, owing to the secondary circuit being
broken before the current in it was fully formed.

The following deflections were obtained :—

Turns per second.						Deflection.
10·10	.	.	.	.	.	. 38·34
12·23	.	.	.	.	.	. 7
16·66	.	.	.	.	.	. 1
19·05	.	.	.	.	.	. ½
19·90, and above		.	.	.	.	0

That is, that when the velocity was greater than 19·9 turns per second, the current was broken before the induction current began to be formed. From these experiments we obtain the following results :—

"*The induced* current is *not formed* suddenly *at the moment of closing, but after a certain interval of time. It arrives rapidly at a maximum, and then decreases more slowly.*"

Professor Blaserna states that the shape of the curve expressing the relation between the velocity of revolution and the strength of the induced current varies, not only with the distance apart of the spirals, but with the specific inductive capacity* of any substance which is placed between them.

The time t which elapses between the closing of the primary and the commencement of the formation of the induced current, depends on the distance between the spirals and the substance between them.

The following are some of Blaserna's results :—

Distance between the Spirals.	t
1 centim. of air	0·000167 seconds
2·3 „ „ „	0·000208 „
2·3 with disc of shellac . . .	0·000380 „
4 „ „ of air	0·000290 „
1 centim. with large disc of shellac .	0·000450 „
1 centim. with 4 layers of glass, each of thickness of shellac disc . . .	0·000373 „
1 centim. with disc of sulphur . .	0·000402 „
1 centim. with disc of resinous pitch †	0·000585 (?) „

From these and similar experiments he deduced the distance which there would have to have been between the spirals, in order that the induced current should commence one second after the closing of the primary when the space between was filled with different substances.

---

* More probably the magnetic inductive capacity. Experiments with the Hughes Induction Balance, vol. i. p. 298, might elucidate this point.—J.E.H.G.

† Pece greca.

We have

for Air . . . . 270 metres
       Glass . . . . 61 ,,
       Shellac . . . . . . 57 ,,
       (Large disc) . . . . . 44 ,,
       Sulphur . . . . . 52 ,,
       Resinous pitch* . . . . . 30 (?) ,,

These distances he calls the " retardation" produced by the different substances.

END OF THE INDUCED CURRENT.

For this portion of the investigation the method adopted was, to so arrange the cylinders that the secondary current was closed about 3° *after* the primary; so if the phenomena were not complicated ·by any disturbing cause, the duration of the induced current, or rather the interval between the closing of the primary and the end of the induced current, would be the time taken by the cylinder to turn 3° when the speed was the least that gave no deflection of the galvanometer.

The phenomenon is, however, complicated by two things.

The reaction of the secondary current on the primary circuit, and the doubt which exists as to whether or not the induced current ceases at the moment of breaking the circuit or whether it remains for a short time in the wire.

Blaserna was not able fully to correct for these errors, but he gives as an approximation, that the duration of the current of closing varies from 0·001624 second to 0·001960 second. He says, however, that he attaches but little value to these results. He considers it possible that there may be several maxima and minima.

THE INDUCED CURRENT OF OPENING.

For experiments on this the cylinders were arranged so that the secondary current was closed at the moment when, or for greater precaution a moment before, the primary was opened.

It was then found that—

The current of opening is both formed and completed in a time shorter than that of the current of closing.

The following comparative results are given :—

Time required to attain maximum under the same conditions,

       Current of closing . . . . 0·000485 seconds.
       Current of opening . . . 0·000275 ,,

---

* Pece greca.

Maximum intensity,

Current of closing	.	.	.	23770 arbitrary units.
Current of opening	.	.	.	86000 „ „

It must be remembered, however, that the "area" of the two currents is identical; that is, that the time integral* has the same value for each.

This means that the total quantity of electricity carried across the wire is the same in each direction.

Then, if the current of opening is more intense, and less in duration than that of closing, the difference of potential along the secondary must be greater in the former case than in the latter.

Professor Blaserna finds that the ratio of the potentials in the two cases is about 13 to 6.

Now, Messrs. De La Rue† and Müller have found that the length of the spark obtained in air from batteries of 600 to 5640 cells varies as the square of the number of cells; that is, as the square of the difference of potential.

Now, suppose the induced current to be strong enough to give sparks, we should by Blaserna's results have—

Ratio of length of spark due to current of opening to that due to current of closing,

$$= 13^2 : 6^2 = 4.69.$$

But in addition to having greater intensity, the current of opening is formed much more quickly than that of closing. Now, the effect of the correction which has been applied to the numbers

$$860, 237$$

to obtain the ratio 13 : 6 appears to me, as far as I understand Professor Blaserna's reasoning, to be to almost eliminate this fact, which is known to be very important to the length of spark.

Taking the uncorrected numbers, we have the

Ratio $860^2 : 237^2 = 13$ about.

In an induction coil ‡ the spark produced on opening is very much greater than that produced on closing; much more than thirteen times greater.

---

* See note † to page 330.
† *Proc. Roy. Soc.*, 1876, vol. xxiv. p. 167.
‡ See vol. ii. ch. xxxix.

z

The cases are, however, hardly comparable, as in the induction coil the effects are complicated by the time required to magnetize and demagnetize the iron core.

The uncorrected value of the ratio, which is simply taken from the swing, is to a certain extent analogous to the spark length, as the swing depends not only on the strength of the impulse but on its suddenness.

The Extra Current of Closing.

For the investigation of the extra current one coil S was used, and it was connected as shown in fig. 150.

A short band being used and the wheel not being turned too fast, oscillations were observed in the current which showed that

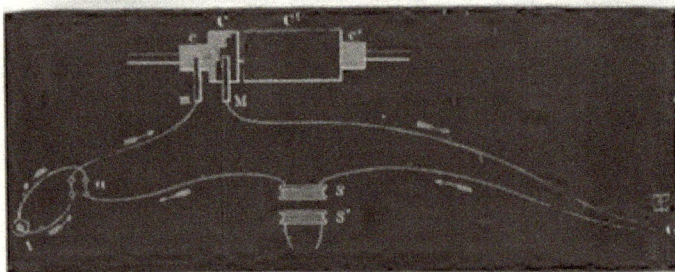

Fig. 150.

the extra current acted against the primary. The first result obtained was that—

"*The induced current forms itself together with the principal current, and commences to circulate (however feebly) suddenly at the moment of closing.*"

Further observation shows that there are several oscillations, that is, that instead of the extra current attaining a maximum and then diminishing, it passes through several maxima and minima.

In Plate XXVII., curves v. and vi. represent the values of the primary current, minus the extra current; and, therefore, measuring the ordinates downward from the top gives the values of the extra current.

The variable state of the extra current may last as much as $\frac{2}{100}$ of a second.

A study of the curves shows that—

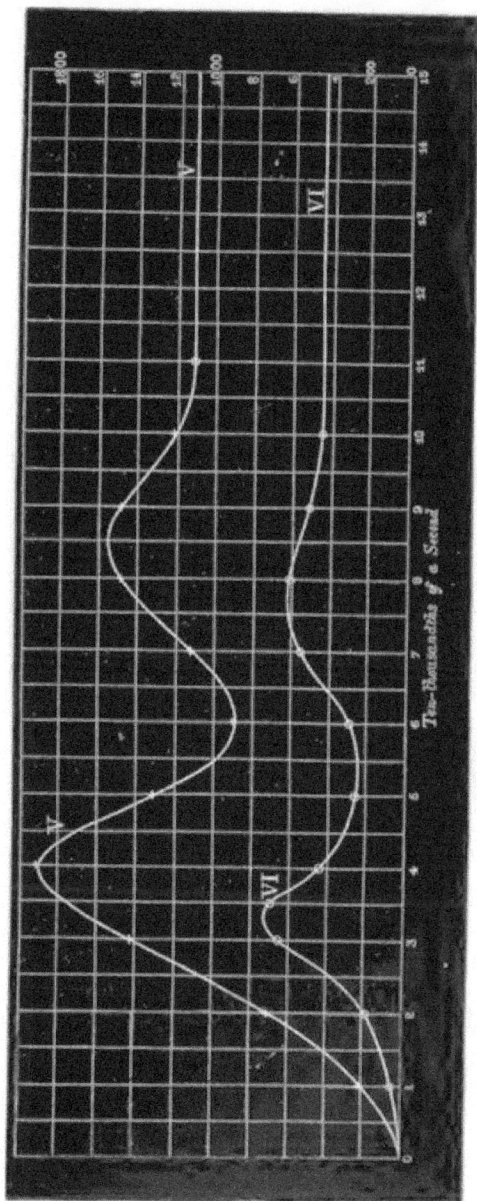

PLATE XXVII.—EXTRA CURRENT OSCILLATIONS.

" *The first oscillation of the current has a great amplitude and
a small duration ; the second, less amplitude and more duration ;
that the amplitude goes on regularly diminishing and the duration increasing, till at last all the oscillation vanishes and is
confounded with the horizontal straight line, which represents the
normal state of the current.*"

We may account for the oscillations in this way :—The extra
current always tends to alter the direction in which the current
is varying. Thus, while the current is increasing, the extra current tends to make it decrease ; as soon as it begins to decrease,
the extra current tends to make it increase, and so on.*

THE EXTRA CURRENT OF OPENING.

This is much more difficult to study than the current of closing,
as an arrangement of the interruptor has to be made, such that
at the moment of opening the battery-circuit, the coil shall be
connected with the galvanometer.

Having made this arrangement Blaserna found, first, that—

" *The extra current of opening consists of alternating currents
which succeed each other at short intervals of time.*"

This was the result of some preliminary experiments; further
observations showed that—

" *The extra current of opening consists of oscillations, more or
less energetic, which are much more rapid than those of closing,
and of which the duration is much shorter.*"

LAW OF LENZ.

In 1834 Lenz enunciated this law—

" If a constant current flows in the primary circuit A, and if by
the motion of A, or of the secondary circuit B, a current is induced
in B, the direction of the induced current will be such that, by its
electro-magnetic action on A, it tends to oppose the motion of
the circuits." †

If, for a current moving away from, or towards another, we
substitute a current whose strength is diminishing or increasing,
we shall see that Blaserna's results, particularly as to the oscillation of the extra current, can be explained as deductions from
Lenz's law.

* This explanation is a mere deduction from Lenz's law.
† *Maxwell's Electricity,* 542, vol. ii. p. 176.

Schiller's Experiments on Electric Oscillations.[*]

In 1874, Schiller published an account of some experiments on "Electric Oscillations." He found that if an electric current be suddenly interrupted in a primary coil R_1, which acts by induction on a secondary coil R_2, *the ends of R_2 being insulated from each other*, that there will be a great number of extremely rapid "oscillations" in the secondary; that is, that if one end of the secondary be connected to the earth, the other will have rapidly alternating (+) and (−) potentials.

The duration of each oscillation varied in different experiments, but generally it was from 6 to 12 hundred-thousandths of a second.

When the ends of the secondary were connected to a condenser the length of each oscillation was increased.

The oscillations were measured by the following method :—

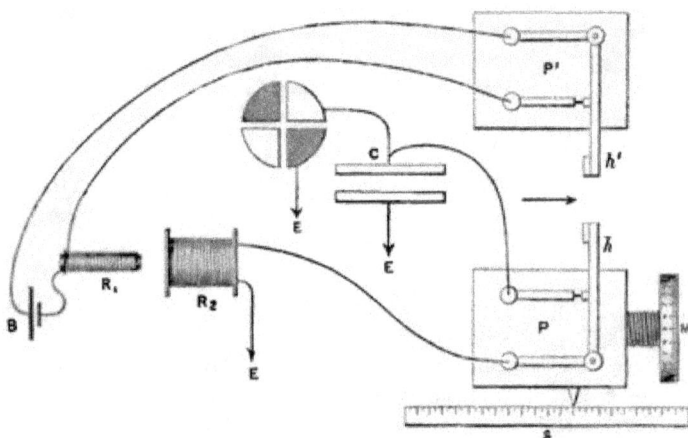

Fig. 161.

A primary coil R_1 was connected to one Daniell cell B, and to a contact key P'.

R_1 was placed inside a secondary coil R_2, one end of which was connected to earth, and the other through a key P to a condenser C, and to one pair of quadrants of an electrometer. The other

[*] Einige experimentelle Untersuchungen über electrische Schwingungen von N. Schiller, Pogg. Ann. 1874, p. 535.

quadrants and the other plate of the electrometer were connected to earth.

The keys P and P′ were ordinarily kept closed by springs, but were so arranged that when opened they would remain open.

The keys were opened by means of a heavy pendulum, which was kept in a horizontal position by means of an electro-magnet, and, on being released, fell in the direction of the arrow, and struck the levers *l l′*.

When *l* and *l′* were in the same straight line, the contacts were broken simultaneously.

The key P could, however, be moved by means of the micrometer screw M, so that the lever *l* was not struck till a certain time after the lever *l′*. The scale S showed the number of turns of the screw, and fractions of a turn were read on the divided head M.

When the velocity of motion of the pendulum was known, the time interval corresponding to one division of the scale S could be determined.*

Thus it was possible to break the contact in the secondary coil, at a small known interval of time after that of the primary.

When the electric oscillations take place, the potential becomes alternately (+) and (−).

The curve of oscillation is of the same general kind as the curve represented by that part of curve V., Plate XXVII., which lies between the verticals 2 and 11, if the horizontal line 13 be taken as the line of zero potential.

Now, as the potential is changing from (+) to (−) or from (−) to (+), it passes through the zero line, and thus, at the beginning and end of each oscillation, the potential in the secondary wire of the electrometer will be zero.

If the secondary circuit is broken at the commencement of an oscillation, there will be no deflection of the electrometer, for, at the moment of breaking, the potential is zero, and the sums of the preceding positive and negative charges are equal.

In the experiments the primary was first broken, and then,

---

* The time value of one scale division was determined by connecting the two keys to the coils of a delicate differential galvanometer. When the contacts were broken simultaneously, there was no deflection. When P was moved one scale division, there was a swing of the needle depending on the time during which the current in P had acted alone. The time was calculated from the swing.

after a convenient interval, **the secondary** was broken. The interval was chosen so that **the results should not be** confused by the spark, but yet that the oscillations should not be too small.

The micrometer screw **was then** adjusted till there was no deflection of the electrometer.

The reading having **been taken, the key P** was screwed on to the next position, in which **there was no deflection.** The difference between the first **and second** readings gave the duration of an oscillation in scale divisions. **The** time corresponding to one scale division was determined by separate experiments.

For the particular instrument used, it was found **to be**

$$\cdot 0000012536 \text{ sec.}$$

Without any condenser, the duration of an oscillation was found to be 33 scale divisions, or

$$\cdot 000041369 \text{ sec.}$$

(four one hundred thousandths of a second).

The author shows how to allow for the "damping effect" caused by the induction of the wires on each other, &c.

Application to Specific Inductive Capacity.

It can be shown mathematically that the capacity of a coil is proportional to the square of the time of oscillation when the coil only is used, and that of a coil and condenser together to the same quantity when the coil and condenser are used.

From this fact the ratio of the capacities of two condensers can be calculated.

Now, if we have two exactly similar condensers—one containing air, and the other some dielectric—then the ratio of the capacity of the second to that of the first is the specific inductive capacity of the dielectric.

Let T_0 be the oscillation time with coil only,

T　,, ,,　　,,　　,, with coil and dielectric condenser.

T'　,, ,,　　,,　　,, with coil and same condenser, with dielectric removed, and only air in it;

then we shall have for the specific inductive capacity

$$K = \frac{T^2 - T_0^2}{T'^2 - T_0^2}.$$

A series of determinations of specific inductive capacity were made by this method with the results given on page 103. They are of particular importance, as, owing to the rapidity of the reversals, there could have been no permanent charging, and the number obtained must represent what Wüllner[*] calls the "Instantaneous capacities of the dielectrics."

* Vol. i. page 106.

END OF VOL. I.

LONDON :
GILBERT AND RIVINGTON, LIMITED,
ST. JOHN'S SQUARE.

www.ingramcontent.com/pod-product-compliance
Lightning Source LLC
Chambersburg PA
CBHW021347210326
41599CB00011B/779